Fault Trees

Fault Trees

Nikolaos Limnios

First published in France in 2005 by Hermès Science/Lavoisier entitled "Arbres de défaillances"
First Published in Great Britain and the United States in 2007 by ISTE Ltd

ISTE Ltd
6 Fitzroy Square
London W1T 5DX
UK

ISTE USA
4308 Patrice Road
Newport Beach, CA 92663
USA

www.iste.co.uk

Library of Congress Cataloging-in-Publication Data

Limnios, Nikolaos.
[Arbres de défaillances. English]
Fault trees/Nikolaos Limnios.
p. cm.
Includes bibliographical references and index.
ISBN-13: 978-1-905209-30-9
ISBN-10: 1-905209-30-4
1. Reliability (Engineering) 2. Trees (Graph theory) I. Title.
TA169.L555 2006
620'.00452--dc22

2006033027

British Library Cataloguing-in-Publication Data
A CIP record for this book is available from the British Library
ISBN 10: 1-905209-30-4
ISBN 13: 978-1-905209-30-9

Printed and bound in Great Britain by Antony Rowe Ltd, Chippenham, Wiltshire.

Table of Contents

Introduction

Dependability: a generic term encompassing the concepts of reliability, availability, maintainability, security, etc. It is also simply designated by the term "reliability", which underlines its quantitative aspect.

The aim of reliability is the study of systems (sets of elements – hardware, software, human resources – that interact with a view to accomplishing a mission) that are subjected to physical processes such as the processes of failure, repair, and stresses.

The component is a part of a system which is non-resolvable within the framework of the study for which sufficient qualitative information (functioning, modes of failure, etc.) as well as quantitative information (failure rate, repair rate, etc.) has been provided in this study. The notion of component is relative and depends on the study. For example, an aircraft, within the framework of a study dealing with flight safety, represents the system, whereas for the airline company it represents a component.

The study of failures in components has led to very elaborate classifications. The failures that have been taken into account in this book are catastrophic failures, that is to say, they are sudden and complete.

As for reliability, the part that deals with the modeling of the components with a view to obtaining qualitative and quantitative information is called "component reliability" or "reliability statistics". Another part of reliability, called the "reliability of systems", is concerned with the modeling of systems with a view to studying their reliability according to the reliability of their components. The reliability of the systems and the component reliability can be complementary in that the results of the former form the data for the latter.

In this book, we will consider diverse classes of systems: in general, many criteria such as those regarding the number of system components with single- or multi-components, the system's structure function presenting a certain form of coherence or non-coherence, state spaces (system and components), binary systems or multi-performance systems, maintenance systems that are non-repairable, reliably repairable, repairable, the mission expressed by a structure function or not have been used.

Considerable diversity exists among reliability models. Excluding the diverse theories, prolongation and applications, we will be considering two large families, namely, models of minimal sets (cut sets and minimal paths) and models involving stochastic processes. The former, not possessing the accuracy and analytical comfort of the latter, have the advantage of considerably reducing the size of problems and of enabling, in most cases, their resolution.

From an algorithmic viewpoint, the complexity of systems in terms of reliability is generally determined by different elements such as a large number of components, a non-classical structure, the existence of certain forms of non-coherence, many levels of performance, extensive variables, non-constant hazard rates, stochastic dependences, and the coexistence of the three elements that is hardware-software-human factor. The problem of evaluating the reliability of a system is an NP-difficult problem ([ROS 75]), that is to say, there is no algorithm whose time execution is limited by a polynomial function of the problem's size (i.e., number of components), unless, for any class of problems considered as being difficult, there exists a polynomial algorithm.

The main problem of reliability is the construction of the structure function and the probabilistic risk assessment.

Fault trees: the fault tree (FT) forms part of the family of models called minimal sets, that is, the models using the minimal cut sets and/or the minimal paths for studying the reliability of systems. It was developed with the aim of making it possible to obtain the cut sets of complex systems. At present, the FT constitutes one of the most widely employed methods in the domain of the reliability of systems.

Designed by Watson 1962 in the laboratories of the "Bell Telephone Company" and within the framework of the project involving the "ICBM minuteman" missiles ordered by the US Air Force, it saw three stages of development. Initially, during the 1960s, it served as a tool for representing system failures but in the absence of the techniques and algorithms that are specific for its treatment. Subsequently, Haasl introduced the basic rules for the construction of the FTs in 1965, Vesely in 1970 [VES 70] supplied us with the "Kinetic Tree

Theory" called Kitt, where, through the underlying stochastic processes, the design of the FTs has become more complete; this theory remains the main tool for the quantitative evaluation of the FTs until now. At the same time, Fussell and Vesely [FUS 72] developed and perfected the MOCUS algorithm, which is distinct from the algorithms of combinatorial character. The third stage of development, in the 1980s, was marked by the extension of this theory to the non-coherent fault trees, multistate fault trees and fuzzy fault trees.

Recently, a new algorithm has considerably improved the calculation performances and has offered the possibility of large FTs; the algorithm described in this study is based, on the one hand, on recursive algorithms that do not require prior information of the minimal sets of the FT and, on the other hand, on the truncation algorithms of minimal cut sets.

The FT is a purely deductive technique. An FT represents a failure mode of a system according to the failure modes of its subsystems and components. The term "fault tree" is to a certain extent restrictive; for example, we will go on to discuss a dual FT which, in principle, represents the good functioning of a system (in the case of binary systems) is described in this study. Barlow and Proschan [BAR 75] make use of the term "event tree" and not "fault tree") for designating an FT; this term also adds to the ambiguity, for it also designates the inductive event trees [LIM 84]. For distinguishing it from the latter, we could use the term "deductive events tree". Nevertheless, in this book, we will be focusing on the use of the traditional term of "fault tree"; however, in a number of cases, it will not represent the failure but the good functioning of the systems.

Figure 1 shows the essential stages for the evaluation of the reliability of a system (1-4-5), that is, proceeding from the system, we obtain its structure function that we will introduce in a model of probabilities for evaluating its reliability. Obtaining the function of structure from the system is a difficult task and, except in the case of simple systems (in principle, systems of elementary structure), this cannot be done without special tools for the majority of complex systems. Thus, modeling of the system is obtained through standard graphs, of which FT is part, for obtaining in a systematic manner the structure function. As a result, the FT is employed right from the first stages of safety analysis for the functioning of the systems. The safety study of a system through FT includes three stages: the construction of the tree, qualitative analysis and quantitative analysis. This construction should be highly exhaustive, that is, representing all the (significant) causes for the failure of the system. The construction technique can be obtained quit quckly, which greatly facilitates the collaboration of specialists of diverse domains.

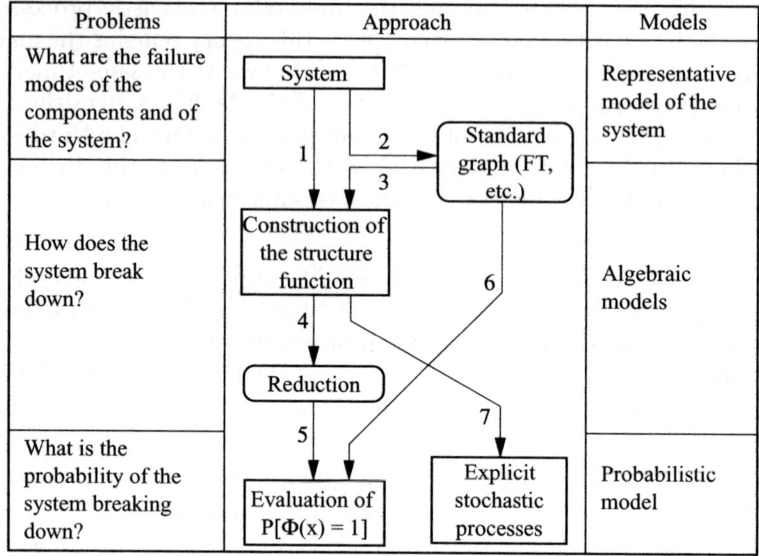

Figure 1 *Problems, approach and models for evaluating the reliability of systems*

The qualitative analysis deals with methods for obtaining the minimal sets: minimal cut sets and minimal paths. The quantitative analysis comprises on the one hand the evaluation of the probability of the top event (within the framework of the preliminary analysis of the risks, this event is called "undesirable") and on the other hand the study of influence concerning the sensitivity and the importance of the basic events vis-à-vis the top event. The evaluation of the probability of the top event can be carried out directly on the FT without passing through the minimal sets, when the FT does not contain repeated events. Another use for the FT, particularly for its minimal cut sets, is concerned with the division of the spaces of states into states of running and into states of breakdown of the systems modeled by the stochastic processes.

The undisputed efficacy of FTs for representing failures of complex systems encounters difficulties when probabilistic treatment is concerned. This is a limitation that is common to the methods based on minimal sets and is linked to the two following impossibilities: one representing the exact calculation of the reliability for the systems with repairable components and the other concerned with the calculation in case of certain dependences. In actual practice, we overcome this limitation by making an approximate calculation, which in the case of the systems of a good reliability is having the correct accuracy.

The FT is used at first for analyzing the failures of the hardware and then for modelizing human failures or errors [DHI 86]. Its use is still very much limited in the software domain [LEV 83]. In principle, the FT can contain events concerning the software but is used very rarely for analyzing a software independent of its application.

Organization of the book: the FTs are at first presented for modeling the coherent binary systems, we refer to them as coherent FTs (c-FTs). Then, we face certain extensions such as the non-coherent FTs (nc-FTs) and the FTs with restrictions (FT-r), which represent a generalization of the nc-FT and the multistate FTs (m-FTs).

Before FTs are described, it is important to present in Chapters 1 and 2 the basic elements necessary for the study of FTs. In Chapters 3–9, FTs are discussed. In Chapter 10, the elements of stochastic simulation for FTs will be presented.

Chapter 1 deals with the basic relationships concerning the reliability of the binary component, wherein the notions of reliability, availability, maintainability, MTTF, etc., are introduced and expressed through their analytical expressions.

Chapter 2 presents the structure function, which will form the basis for the later development, the diverse classes of systems (systems with elementary structure, systems with complex structure, etc.), the reliability function and the general methods of evaluation.

Chapter 3 deals with the construction of FTs: the different graphic symbols and the stages of construction.

Chapter 4 covers qualitative treatment, that is, the search for minimal sets, and also the corresponding classical algorithms.

Chapter 5 deals with quantitative treatment: the diverse methods of evaluating the probability of the top event and the essential methods for the evaluation of large FTs.

Chapter 6 gives a study of influence: the uncertainty or the sensitivity and the importance, followed by the methods of calculating the uncertainty and the most well-known factors of importance.

Chapter 7 deals with the modularization of FTs, multi-phase FTs and the treatment of common failure modes.

Chapter 8 presents certain extensions: non-coherent FTs, delayed action FTs, FTs with restrictions and multistate FTs.

Chapter 9 presents new algorithms based on the binary decision diagrams (BDD).

Chapter 10 presents the stochastic simulation (or Monte Carlo method) for the evaluation of the probability of the top event and other quantities.

In this book, for designating the different parts of a system, apart from the notions of "system" and "component", the notion of the "sub-system" is used; it designates a part of a system containing at least one component and is endowed with a sub-mission within the framework of the overall aim. For designating without any special discrimination, a system, a subsystem or a component, we make use of the notion of "item".

Chapter 1

Single-Component Systems

1.1 Distribution of failure and reliability

1.1.1 *Function of distribution and density of failure*

We will study here the stochastic behaviour of single-component systems being subjected to failures (breakdowns) by observing them over a period of time. Let us simplify things by assuming that the system is put to work at the instant $t = 0$ for the first time and that it presents a single mode of failure.

The component, starting a lifetime period at the instant $t = 0$, is functioning for a certain period of time X_1 (random) at the end of which it breaks down. It remains in this state for a period of time Y_1 (random) during its replacement (or repair) and, at the end of this time, the component is again put to work and so on. In this case, the system is said to be repairable. In the contrary case, that is to say, when the component breaks down and continues to remain in this state, the system is said to be non-repairable.

It is possible to present a graphic description of the behavior of the above-described system in different ways, the phase diagram being the most common.

Let X be a random variable (r.v.) representing the lifetime of the system with F, its cumulative distribution function (c.d.f.):

$$F(t) = P(X \leq t).$$

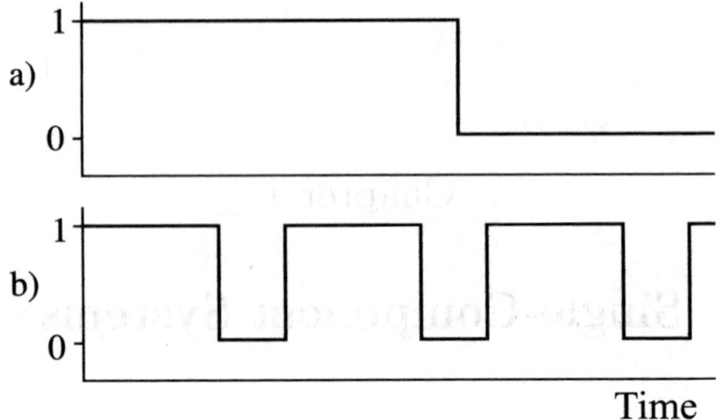

Figure 1.1 *Phase diagrams: (a) non-repairable system and (b) repairable system 1: state of good functioning 0: state of breakdown*

If F is absolutely continuous, the random variable X has a probability density function (p.d.f.) f and can be written as:

$$f(t) = \frac{d}{dt}F(t) = \lim_{\Delta t \to 0} \frac{P(t < X \leq t + \Delta t)}{\Delta t}.$$

Regarding the probability evaluation of fault trees, we always have to make the distinction between the occurrence or arrival of an event and its existence at the time t. Let us consider, for example, that the f.r. F of the duration of life of a component has an p.d.f. f. The assertion "the occurrence of the failure of the component at the time t" means that the failure took place within the time interval $(t, t + \Delta t]$, where $\Delta t \to 0$; as a result, its probability is given by: $f(t)\Delta t + o(\Delta t)$. On the other hand, the assertion "existence of the failure at the time t" means that the failure took place at the time $x \leq t$ and its probability is then simply $F(t)$.

1.1.2 *Survival function: reliability*

The complementary function of F, noted as $\overline{F}$, is called the survival function or reliability of the system, noted as $R(t)$. That is to say:

$$R(t) = \overline{F}(t) = 1 - F(t) = P(X > t),$$

Now

$$R(t) = \int_t^{\infty} f(u)du,$$

Thus we have:

$$R(0)=1, \quad R(+\infty)=0.$$

1.1.3 *Hazard rate*

The hazard rate function, noted as $h(t)$, plays a leading role in connection with the reliability of systems. In the case of a failure process, it is called the failure rate (instantaneous), noted as $\lambda(t)$, and in the case of a process of repair, it is called the (instantaneous) repair rate, noted as $\mu(t)$. In survival analysis, it is also called the risk rate. It is defined as follows:

$$h(t)=\lim_{\Delta t\to 0}\frac{P(t<X\leq t+\Delta t|X>t)}{\Delta t}.$$

Properties:

$$h(t)\geq 0,$$

$$\int_0^\infty h(u)du=+\infty.$$

The cumulative hazard rate $H(t)$ is defined by the relationship:

$$H(t)=\int_0^t h(u)du.$$

The total hazard rate is defined by:

$$H=\int_0^X h(u)du.$$

H follows an exponential distribution of parameter 1.

1.1.4 *Maintainability*

The maintainability, noted as $M(t)$, is defined by the probability that the system will be repaired within the time interval $(0,t]$, given that it broke down at the instant $t=0$. Let Y be the random variable indicating the duration of breakdown (or duration of repair) of the component, and G its distribution function:

$$G(t)=P(Y\leq t).$$

If G is completely continuous, Y has a density noted as g and we will have:

$$g(t)=\frac{d}{dt}G(t).$$

The function G is called as maintainability noted as $M(t)$. The maintainability $M(t)$ is defined by the probability that the system will be repaired within the time interval $(0,t]$, given that it broke down at the instant $t=0$. The repair rate, noted as $\mu(t)$, will be given by:

$$\mu(t) = \lim_{\Delta t \to 0} \frac{P(t < Y \leq t + \Delta t \mid Y > t)}{\Delta t}.$$

1.1.5 *Mean times*

The following mean times (when they exist) play a very important role in connection with the reliability, because they constitute the indices for comparing the reliabilities of systems and of the çomponents supplied generally by the manufacturers.

Mean time to failure (MTTF):

$$MTTF = E[X] = \int_0^{\infty} t dF(t) = \int_0^{t} t f(t) dt.$$

Mean time to repair (MTTR):

$$MTTR = E[Y] = \int_0^{t} t dG(t) = \int_0^{t} t g(t) dt.$$

Mean duration of good functioning after the repair, MUT (Mean Up Time).

Mean duration of failure, MDT (Mean Down Time).

Mean time between failures (MTBF):

$$MTBF = MUT + MDT.$$

In order that MTTF should exist, it is necessary that $\xi > 0$, such that:

$$\lim_{t \to +\infty} e^{\xi t} R(t) = 0.$$

▷ **Example 1.1.** If the density f of the random variable X, with non-negative values, is given by the formula (Cauchy distribution):

$$f(x) = \frac{2}{\pi} \frac{a}{x^2 + a^2}, \quad (a > 0),$$

then X does not have any moment!

It should also be noted that for the single-component systems we will have:

$$MUT = MTTF \quad \text{and} \quad MDT = MTTR,$$

Hence, the formula $MTBF = MTTF + MTTR$ is applicable to this case only.

1.1.6 *Mean residual lifetime*

It will be of interest to know the residual lifetime of the system at the age of t, knowing that it did not break down in $(0,t)$:

$$L(t) = E[X - t|X > t],$$

$$L(t) = \int_t^{\infty} \frac{R(u)du}{R(t)}.$$

This function L(t) satisfies the following conditions:

$$\begin{aligned}
&L(X) = 0,\\
&L(0) = E(X),\\
&\frac{d}{dt}L(t) \geq -1,\\
&\int_0^{\infty} \frac{dt}{L(t)} = +\infty.
\end{aligned}$$

1.1.7 *Fundamental relationships*

The reliability verifies the following relationships:

$$\frac{d}{dt}R(t) + \lambda(t)R(t) = 0,$$

obtained by the definition of the failure rate in section 1.1.3. The solution for this equation is:

$$R(t) = R(0)\exp(-\int_0^t \lambda(u)du).$$

Agreeing that the reliability at the instant $t = 0$ is equal to 1, this relationship can be written as:

$$R(t) = \exp(-\int_0^t \lambda(u)du).$$

This is the most general relationship for the reliability.

In the same manner, we can obtain the equation verified by the maintainability:

$$\frac{d}{dt}M(t) - [1 - M(t)]\mu(t) = 0.$$

The general solution is as follows:

$$M(t) = 1 - [1 - M(0)]\exp(-\int_0^t \mu(u)du).$$

Agreeing that for the maintainability $M(0) = 0$, the preceding relationship becomes:

$$M(t) = 1 - \exp(-\int_0^t \mu(u)du).$$

For the mean times and when the functions R(t) and 1-M(t) are summable over the real half line $x \geq 0$, we can write:

$$MTTF = \int_0^\infty R(t)dt,$$

$$MTTR = \int_0^\infty [1 - M(t)]dt.$$

▷ **Example 1.2.** An electrical equipment has a constant failure rate λ. The reliability of the apparatus at time t is given by $R(t) = \exp(-\lambda t)$. The probability that it might break down in the time interval $(t_1, t_2]$, $(t_1 < t_2)$ is given by:

$$P(t_1 < T \leq t_2) = \int_{t_1}^{t_2} f(t)dt = e^{-t_1} - e^{-t_2}.$$

The probability that it might survive at the time t_2, given that it did not break down between 0 and t_1, is

$$P(T > t_2 | T > t_1) = \frac{P(T > t_1, T > t_2)}{P(T > t_1)} = \frac{P(T > t_2)}{P(T > t_1)} = e^{-\lambda(t_1 - t_2)}$$

and $MTTF = \int_0^\infty e^{-\lambda t} dt = 1/\lambda$.

1.1.8 *Some probability distributions*

We present here some of the usual probability distributions dealing with the reliability of systems that we are using or with those that we will be referring to in the wake of our statement. These probability distributions, unless otherwise mentioned, have their support on the real half-line $x \geq 0$.

– **Exponential distribution**

The exponential distribution is by far the most frequently used in relation to the reliability of systems. A system whose stochastic behavior is modeled by an exponential distribution is a system without memory or a Markovian system, that is to say, for $t > 0$, $x > 0$, we have $P(X > t + x|X > t) = P(X > x)$. For the exponential distribution, we have, for $x \geq 0$:

$$f(t) = \lambda e^{-\lambda t},$$

$$R(t) = e^{-\lambda t},$$

$$\lambda(t) = \lambda.$$

This distribution gives good modeling for the lifetime of electronic components. Nevertheless, its use in other fields, such as for the modeling of mechanical components or the times to repair, is not always justified.

– **Normal distribution**

The normal distribution is supported by the complete real line; as a result, it is not suited for modeling the system's lifetime. Nevertheless, when $0 < \sigma/\mu \ll 1$, the part of the distribution carried by the half-line $x < 0$ can be neglected and the distribution is used in this case for modeling the duration of the lifetime of the systems.

$$f(t) = \frac{1}{\sqrt{2\pi}\sigma} e^{-\frac{(t-\mu)^2}{2\sigma^2}},$$

$$R(t) = \int_t^{\infty} f(y)dy,$$

$$\lambda(t) = \frac{f(t)}{R(t)},$$

where μ is the average and σ is the standard deviation.

– **Log normal distribution**

The log normal distribution models the times of repair quite well. It is also used for the modeling for the propagation of uncertainties in the fault trees.

$$f(t) = \frac{1}{\sqrt{2\pi}\sigma t} e^{-\frac{(\ln t-\mu)^2}{2\sigma^2}}, \quad t \geq 0,$$

$$R(t) = \int_t^{\infty} f(y)dy,$$

$$\lambda(t) = \frac{f(t)}{R(t)},$$

where μ is the average and σ is the standard deviation.

– **Weibull distribution**

Thanks to the vast variations of form that it can take up according to the values of its parameters, the Weibull distribution is used in many domains of reliability, particularly in those concerned with the reliability of mechanical components.

$$f(t) = \frac{\beta}{\eta^\beta}(t-\gamma)^{\beta-1}\exp\{-\frac{(t-\gamma)^\beta}{\eta}\},$$

$$R(t) = \exp\{-\frac{(t-\gamma)^\beta}{\eta}\},$$

$$\lambda(t) = \frac{\beta(t-\gamma)^{\beta-1}}{\eta^\beta},$$

where β is the parameter of form, η the parameter of scale and γ the parameter of localization. For $\beta = 1$ and $\gamma = 0$, we will obtain the exponential distribution.

– **Gamma distribution**

The gamma distribution has, just like the preceding distribution, multiple forms depending on the values of its parameters.

$$f(t) = \frac{(\lambda t)^{a-1}}{\Gamma(a)}\lambda e^{-\lambda t},$$

$$R(t) = \frac{\lambda^a}{\Gamma(a)}\int_t^\infty y^{a-1}e^{-\lambda y}dy,$$

$$\lambda(t) = \frac{f(t)}{R(t)},$$

where $\lambda > 0$ and $\Gamma(x) = \int_0^\infty t^{x-1}e^{-t}dt$ is the gamma function. For $n \in \mathbb{N}^*$, we have $\Gamma(n) = (n-1)$.

If $a = 1$, we are then dealing with the exponential distribution. If $a \in \mathbb{N}^*$, then the gamma distribution is called the Erlang distribution and the reliability is given by the following relationship:

$$R(t) = e^{-\lambda t}\sum_{k=0}^{a-1}\frac{(\lambda t)^k}{k!}.$$

– **Distribution of the delay systems**

If X and Y are distributed as per the exponential distributions of the parameters λ and μ respectively, and $\tau > 0$, let us consider the random variable T [LIM 90]:

$$T = (X_1 + Y_1^*) + \cdots + (X_{n-1} + Y_{n-1}^*) + X_n + \tau$$

with $n = \inf\{k : \tau \geq Y_i,\ i = 1, ..., k-1, \tau < Y_k\}$ and

$$Y_i^* = \begin{cases} Y_i & \text{if } Y_i \leq \tau \\ +\infty & \text{if } Y_i > \tau. \end{cases}$$

Then the density of probability of T is given by:

$$f_T(t) = \begin{cases} \lambda \sum_{n=0}^{r-1}(\lambda\mu)^n e^{-(n+1)\mu\tau}\Phi_n(t-(n+1)\tau), & \text{if } t \geq \tau, \\ 0, & \text{if } 0 \leq t < \tau, \end{cases}$$

for $r\tau \leq t < (r+1)\tau$, with

$$\Phi_o(t) = \frac{\mu}{\lambda+\mu} + \frac{\lambda}{\lambda+\mu} e^{-(\lambda+\mu)t}$$

and for $n = 1, 2, ...,$

$$\Phi_n(t) = \frac{1}{(n!)^2}\Big\{ n \sum_{k=0}^{n-1} (-1)^{n+k} C_k^{n-1} (1 + \frac{\mu}{n-k} t) t^{n-1-k} I_{n+k}(t) + \mu I_{2n}(t) \Big\}$$

$$I_k(t) = \int_0^t u^k e^{-au} du = \frac{k!}{a^{k+1}} \Big\{ 1 - e^{-at} \sum_{j=0}^{k} \frac{(at)^j}{j!} \Big\}$$

with $t \in \mathbb{R}_+^*$, $k \in \mathbb{N}$, and $a = \lambda + \mu$.

1.2 Availability of the repairable systems

Contrary to reliability, which is concerned with the good working of the system over an interval of time, $[0, t]$, availability deals with the good functioning at the instant t, irrespective of the fact that the system might have had one or more failures prior to t.

In order to study the availability, we will be introducing, for a fixed time of t, the random variable $X(t)$ with the values $\{0, 1\}$:

$$X(t) = \begin{cases} 1 & \text{if the system is in good state at the instant } t \\ 0 & \text{if the system is in down state at the instant } t. \end{cases}$$

1.2.1 *Instantaneous availability*

The instantaneous availability of a system, noted as $A(t)$, is expressed by the probability that the system is in good condition at the instant t.

$$A(t) = P\{X(t) = 1\}.$$

If $X(t), t \geq 0$ is a stochastic process with state space $\{0, 1\}$, the instantaneous availability at the time t is represented by the probability that the process is seen to be state 1 at t. According to the definitions, we will easily establish the following inequality:

$$A(t) \geq R(t), \quad t \geq 0.$$

When the values for the availability are very close to unity, as is the case with standby safety systems, we will rather be using the instantaneous unavailability, noted as $\overline{A}(t)$:

$$\overline{A}(t) = 1 - A(t) = P\{X(t) = 0\}.$$

1.2.2 *Asymptotic availability*

This is expressed by the portion of the period of good working per unit time, when the system is under a "stationary probabilistic condition", that is to say a sufficiently long time after its commissioning.

Mathematically, when it exists, it is expressed by the following limit:

$$A = \lim_{t \to +\infty} A(t).$$

Let $X = (X_i; i = 1, 2, ...)$ represent the times of good working and $Y = (Y_i; i = 1, 2, ...)$ represent the times of breakdown of a system.

If

- X is i.i.d,
- Y is i.i.d,
- X and Y are independent,
- at least one of X and Y is not a lattice,

then stationary availability exists and it is given by the following relationship [ASC 84]:

$$A = \frac{E[X]}{E[X] + E[Y]} = \frac{MTTF}{MTTF + MTTR}.$$

For the same reason as in the preceding case, we will make use of the notion of asymptotic unavailability, noted as $\overline{A}$: $\overline{A} = 1 - A$.

If $X(t)$ is a Markov process with state space $\{0, 1\}$, then $(\overline{A}, A)$ represents the stationary distribution of this process and in this case this distribution always exists since $\lambda > 0$ and $\mu > 0$.

For a non-repairable system, we have: $A = 0$.

1.2.3 *Mean availability*

The mean availability, noted as $\widetilde{A}(t)$, over $[0, t]$ is expressed by the expectation of the time of good functioning of the system over $[0, t]$:

$$\widetilde{A}(t) = \frac{1}{t} \int_0^t A(x) dx.$$

1.2.4 *Asymptotic mean availability*

This is expressed by the following relationship:

$$\widetilde{A} = \lim_{t \to +\infty} \frac{1}{t} \int_0^t A(x)dx.$$

In the case where $\lim_{t \to +\infty} A(t)$ exists, we will have $A = \widetilde{A}$.

The unavailability of a component will be noted as $q(t)$.

– component with constant unavailability:

$$q(t) = q = \text{constant}, \quad t \geq 0.$$

Component with constant failure and repair rates λ and μ respectively:

– non-repairable component:

$$q(t) = 1 - e^{-\lambda t},$$

– repairable component:

$$q(t) = \frac{\lambda}{\lambda + \mu}(1 - e^{-(\lambda+\mu)t}),$$

– asymptotic unavailability:

$$q = \frac{\lambda}{\lambda + \mu} = \frac{\lambda\tau}{1 + \lambda\tau},$$

where τ is the mean repair time ($\tau = 1/\mu$). When $\lambda\tau \ll 1$, the above formula reduces to: $q \cong \lambda\tau$.

1.3 Reliability in discrete time

It will be of interest to calculate the reliability of a system in $\mathbb{N}$ as the calculations are much simpler to carry out. In the section that follows, we will first list the distributions of some discrete random variables and then the reliability over $\mathbb{N}$ [BRA 03].

1.3.1 *Discrete distributions*

Let X be a discrete random variable with values $A = \{a_1, a_2, ..., a_n, ...\}$. Its probability law is defined by the data of a series of probabilities $(p(a_i); i = 1, 2, ...)$ with: $p(a_i) = P(X = a_i), i = 1, 2, ...$ We have:

$$p(a_i) \geq 0, i \in \mathbb{N}^*, \quad \sum_{i=1}^{\infty} p(a_i) = 1.$$

Bernoulli distribution. This is the distribution of random variable X with values $\{0, 1\}$, and is given by:

$$x = 0, 1.$$

Binomial distribution. This is the distribution of random variable X with values $A = \{0, 1, ..., n\}$, and is given by:

$$p(x) = C_x^n p^x (1-p)^{n-x}, \quad x \in A.$$

Geometric distribution. This is the distribution of random variable X with values $\mathbb{N}^*$, and is given by:

$$p(x) = (1-p)^{x-1} p, \quad x \in \mathbb{N}^*.$$

Poisson's distribution. This is the distribution of random variable X with values $\mathbb{N}$, and is given by:

$$p(x) = e^{-\lambda} \frac{\lambda^x}{x!}, \quad x \in \mathbb{N}.$$

1.3.2 *Reliability*

We will give the reliability in $\mathbb{N}$. The failure rate is given as follows:

$$\lambda(n) = P(X = n | X \geq n), \quad (n \in \mathbb{N}).$$

In order that the sequence $(\lambda(n), n \in \mathbb{N})$ represents a failure rate, it is necessary to have

$$0 \leq \lambda(n) \leq 1, \quad (n \in \mathbb{N}),$$

$$\sum_{n \in \mathbb{N}} \lambda(n) = +\infty.$$

Probability distribution of failure at the time n:

$$f(n) = P(X = n) = [1 - \lambda(0)][1 - \lambda(1)]...[1 - \lambda(n-1)]\lambda(n).$$

Reliability:

$$R(n) = P(X \geq n) = [1 - \lambda(0)][1 - \lambda(1)]...[1 - \lambda(n-1)].$$

Availability: if $\lambda(n) = \lambda$ and $\mu(n) = \mu$, for every $n \in \mathbb{N}$, then:

$$A(n) = \frac{\mu}{\lambda + \mu} + \frac{(1 - \lambda - \mu)^n}{\lambda + \mu}(\lambda P_1(0) + \mu P_2(0)),$$

where $P_1(0)$ represents the probability that the component be in good state at the instant 0 and $P_2(0) = 1 - P_1(0)$.

▷ **Example 1.3.** The calculation for the reliability of the electrical apparatus and failure rate, in discrete time, is carried out as follows. We choose a number of intervals n such that: $\Delta t < 1$, where $\Delta t = t/n$. Then, the reliability is given by:

$$R(t) = (1 - \lambda \Delta t)^n = (1 - \frac{\lambda t}{n})^n.$$

This relationship is justified also by the fact that: $(1 - \frac{\lambda t}{n})^n \to e^{-\lambda t}$ when $n \to +\infty$, from where it can be noted that with a smaller Δt, the approximation is better.

1.4 Reliability and maintenance

For reliability, maintenance constitutes a basic given condition. It affects the reliability from both the logical point of view of failure and from the probabilistic point of view. In particular, the maintainability constitutes a measure of the efficacy pertaining to a given maintenance policy.

1.4.1 *Periodic test: repair time is negligible*

Let there be a component for which we are applying a preventive maintenance by replacing it systematically every T hours. We will assume that the time of replacement is negligible.

For $t = jT + \tau$, $j \in \mathbb{N}$, $\tau \in [0, T)$, we have:

– Reliability

$$R(t) = R(jT + \tau) = [R(T)]^j R(\tau).$$

– Availability

$$A(t) = A(jT + \tau) = A(\tau).$$

– Mean time to failure

$$MTTF = \int_0^T \frac{R(u)du}{1 - R(T)}.$$

1.4.2 *Periodic test: repair time is not negligible*

We are considering here the same problem that has been described previously, but allowing for a time of intervention (inspection and/or repair) that is not negligible, that is to say, there are regular interventions at times jT, $j=1,2,\ldots$. During these interventions into the system, there are two possibilities: either the system is working, in which case a time t_1 of inspection is necessary, or it is under breakdown, in which case an additional time t_2 for repair is necessary, and hence the time of stop in this case t_1+t_2. The unavailability is therefore given by the following relationship:

$$q(t) = 1 - \exp\{-(t-jT)\lambda_e\}\{1-\exp[-(\frac{t-jT}{t_1})^a]\},$$

with:

$$\lambda_e = \frac{T-t_1}{T}(\frac{T-t_1}{T}+\frac{2t_2}{T}) + 2(1-\frac{\Gamma(1/a)}{a})\frac{t_1}{T^2}$$

$$a = \ln[3-\ln(\lambda t_1)]$$

j: number of test interval

T: test interval

t_1: time of inspection of a non-defective component

t_2: time of inspection of a defective component

t_1+t_2: time of inspection of a defective component

λ_e: effective failure rate of a component.

1.4.3 *Mean duration of a hidden failure*

Let a component be inspected periodically at times T, $2T$, $3T$, ... Let us consider $t=jT+\tau$ the instant the component's breaks down. We will try to evaluate the duration of the breakdown time before the next inspection: $D=T-\tau$:

$$E(D) = E(T-\tau) = T - \int_0^T F(t)dt.$$

When X follows an exponential distribution of the parameter λ, we have:

$$E(D) = T - \frac{1}{\lambda}\frac{T}{e^{\lambda T}-1} \cong \frac{T}{2}(1+\frac{\lambda T}{6}).$$

1.5 Reliability data

In order to be able to evaluate the reliability of the systems, we should have the database, that is, the data concerning the components: the probabilities of failure and repair or the rates of failure and repair, etc. The sources for the data to obtain the rates of failure of components are the tests and usage. These data can be treated under non-parametric or parametric form for obtaining the desirable measures.

It has to be said that, more often than not, we lack probablistic data. Operators who are devoted to the means for collecting and storing these data are few in number, and those treating them at a sufficient level with existing statistical techniques are even fewer. The majority of the the time, this treatment is limited to getting a constant rate of failure.

Nevertheless, special efforts have been taken up during the last few years to put up the databases for the operational reliability. At the European level, we have the following databases: ERDS (European Reliability Data System, 1980), ODERA (deals with the petroleum platforms, 1980), SRDF (nuclear, EDF, 1978), etc. [AME 85].

The data treatment should contain an exploratory treatment preceding traditional statistical treatment. For example, an exploratory treatment, which consists of tracing a curve in the plane (t, n) (where t is the cumulative time of working and n the number of failures), can supply us with precious information: constant increasing and decreasing rates of failure, etc. [ASC 84], [HIL 90].

1.3 Reliability data

In order to be able to evaluate the reliability of the systems, we should have the statistics, that is, the data concerning the components: the probabilities of failure and the rates of failure and repair, etc. The sources for the data to obtain the rates of failure of components are the tests and usage. These data can be treated under non-parametric or parametric form for obtaining the desirable magnitudes.

It has to be said that, more often than not, we lack probabilistic data. Operators who are devoted to the means for collecting and storing these data are few in number, and those treating them at a sufficient level with existing statistical methods, are even fewer. The majority of the [illegible] is limited to getting a constant rate of failure.

Nevertheless, important efforts have been taken up during the recent years to set up the databases for the operational reliability. At the European level, we have the following databases: EIReDA (European Reliability [illegible] [illegible]), OREDA (deals with the petroleum industry, [illegible]) [illegible] (nuclear, EDF, 1999), etc. [MIE 97].

The data treatment should contain an exploratory treatment preceding the inferential statistical treatment. For example, an exploratory treatment, which consists of tracing a curve in the plane (t, h) where t is the cumulative time of functioning and h the number of failures, can suggest us with a certain approximation, constant, increasing or decreasing rates of failure, etc. [ASC 84, HIL 00].

Chapter 2

Multi-Component Systems

2.1 Structure function

Chapter 1 focused on the reliability of a single-component system. This chapter focuses on the reliability of systems with more than one component, which are referred to as "multi-component systems".

Whereas in the case of single-component systems the failure in the component implies the failure of the system, this is no longer the case for multi-component systems, at least not automatically. The failure in a multi-component system arises in the wake of the failure in the sub-sets of well-defined components. For example, the failure in one of the four cylinders of a car does not jeopardize its overall working: it works badly, but it still works. On the other hand, the failure of the drive shaft leads to the immobilization of the car.

From this example, it can be observed that the sub-sets of the components, whose simultaneous failures lead to the failure of the system, should be defined exactly. This approach makes use of the Boolean algebra, and more specifically uses the structure function, which translates the functional relationships among the components of the system and its state of failure/working.

Let us consider a binary system with n components: $C = \{1, ..., n\}$: For each component i, we define a variable x_i, with value in $\{0, 1\}$, with the following convention:

$$x_i = \begin{cases} 1 \text{ if the component is in good state} \\ 0 \text{ if the component is in down state} \end{cases}$$

Let $\mathbf{x} = (x_1, x_2, ..., x_n) \in \{0,1\}^n$ be the vector describing jointly the states of the components. We define a function $\varphi(\mathbf{x})$ describing the state of the system with values in $\{0,1\}$, with the following convention:

$$\varphi(\mathbf{x}) = \begin{cases} 1 \text{ if the system is in good state} \\ 0 \text{ if the system is in down state} \end{cases}$$

The set $\{0,1\}^n$ fitted with the operations $(\dot{+})$ addition and $(\cdot)$ multiplication is called a Boolean lattice (complemented distributive lattice), which has the common properties of associativity, commutativity and distributivity. The "zero" element is noted as $\mathbf{0} = (0, 0, ..., 0)$, the "unity" element as $\mathbf{1} = (1, 1, ..., 1)$ and the "inverse" element $\overline{\mathbf{x}} = 1 - \mathbf{x} = (1 - x_1, 1 - x_2, ..., 1 - x_n)$. The operations $(\dot{+})$ addition and $(\cdot)$ multiplication among the elements of $\{0,1\}^n$ are carried out as follows:

$$\mathbf{x} \dot{+} \mathbf{y} = (x_1 \dot{+} y_1, x_2 \dot{+} y_2, ..., x_n \dot{+} y_n)$$

$$\mathbf{x} \cdot \mathbf{y} = (x_1 \cdot y_1, x_2 \cdot y_2, ..., x_n \cdot y_n).$$

The set $\{0,1\}^n$ is partially ordered with the relationship "$\leq$" called "inclusion". Strict inclusion "$<$" and equality "$=$" are defined in the usual manner.

▷ **Example 2.1.** Let us consider the brake system of a vehicle, made up of front and rear brakes. If the rear brakes are down, we can still continue braking. On the other hand, if both the brakes are down (front and rear), we can no longer brake and the braking system is declared to be failed. If the front brakes be indicated by x_1 and the rear brakes by x_2, then the braking system of the vehicle can be described by the function:

$$\varphi(\mathbf{x}) = 1 - (1 - x_1)(1 - x_2).$$

The table given below shows the values of this function corresponding to the values of the variables.

x_1	x_2	$\varphi(\mathbf{x})$
0	0	0
0	1	1
1	0	1
1	1	1

The binary operation $(\dot{+})$ can be written as:

$$x \dot{+} y = x + y - xy = 1 - (1 - x)(1 - y) = \max\{x, y\}.$$

The correspondence among the operations utilized is given in the following table:

Sets/ events	Indicator variables
$\cup$	$\dot{+}$
$\cap$	$\cdot$

Denote x_A and x_B respectively, as the indicator functions (Boolean variables) of events A and B. We have the following correspondences:

$$E = A \cap B \Leftrightarrow x_E = x_A x_B$$

$$E = A \cup B \Leftrightarrow x_E = x_A \dot{+} x_B = x_A + x_B - x_A x_B.$$

Remark 2.1. The intersection ($\cap$) will often be neglected or replaced by the comma, when the events are defined by random variables; see section 5.2.

Elementary functions:

$\varphi_1(\mathbf{x}) = 0$	Constant function 0
$\varphi_2(\mathbf{x}) = 1$	Constant function 1
$\varphi_3(x) = x$	Identity function
$\varphi_4(x) = \overline{x} = 1 - x$	"NON"-function (non-x)
$\varphi_5(x_1, x_2) = x_1 x_2$	Conjunction of x_1, x_2
$\varphi_6(x_1, x_2) = x_1 \dot{+} x_2$	Disjunction of x_1, x_2
$\varphi_7(x_1, x_2) = x_1 \rightarrow x_2$	Implication of x_1 by x_2
$\varphi_8(x_1, x_2) = x_1 \oplus x_2$	Or exclusive.

Notations:

$$(1_i, \mathbf{x}) = (x_1, ..., x_{i-1}, 1, x_{i+1}, ..., x_n)$$

$$(0_i, \mathbf{x}) = (x_1, ..., x_{i-1}, 0, x_{i+1}, ..., x_n).$$

Essential variable: the variable x_i is said to be essential if there exists a vector x such as $\varphi(1_i, \mathbf{x}) \neq \varphi(0_i, \mathbf{x})$. On the contrary, the variable x_i is said to be *inessential*.

Equal functions: two functions $\varphi_1(\mathbf{x})$ and $\varphi_2(\mathbf{x})$ are said to be equal if one is deduced from the other or vice versa by adjunction or elimination of the inessential variables. It is noted that: $\varphi_1 \equiv \varphi_2$.

Dual function: given the function $\varphi(\mathbf{x})$, we define its dual function, noted as $\varphi^D(\mathbf{x})$ or $\overline{\varphi}(\mathbf{x})$, as follows:

$$\varphi^D(\mathbf{x}) = 1 - \varphi(\mathbf{1} - \mathbf{x}).$$

Monotone structure function: the function φ is said to be monotone with respect to the variable x_i if:

$$\varphi(1_i, \mathbf{x}) \geq \varphi(0_i, \mathbf{x}), \quad \mathbf{x} \in \{0,1\}^n.$$

If the function φ is monotonic with respect to all the variables, then φ will be called monotonic.

Coherent structure function: if the function φ is monotone and, in addition, all its variables are essential, then φ will be called coherent.

2.2 Modules and modular decomposition

For a sizeable problem, the solution we adopt frequently, not only in the matter of reliability but also in many other domains, is to decompose it, as much as possible, into many small problems in order to be able to study it. In the matter of system reliability, this approach is carried out with the aid of the modules that are defined as follows.

Module: the coherent system (A, α) is a module of the coherent system (C, φ), if:

$$\varphi(\mathbf{x}) = \psi(\alpha(\mathbf{x}^A), \mathbf{x}^{\overline{A}}), \quad \overline{A} = C \setminus A,$$

where $A \subset C$, and $\mathbf{x}^A$ represents the vector with elements x_i,$i \in A$. ψ is a coherent structure function called an *organizing function*, and A is called a modular set of (C, φ).

If $A \subset C$ and $|A| \geq 2$, then (A, ψ) is called a *proper module* of (C, φ). A modular decomposition of the coherent system(C, φ) is a set of disjoint modules:

$$\{(A_1, \alpha_1), ..., (A_r, \alpha_r)\}$$

with an organizing function ψ, if:

(i) $C = \cup_{i=1}^r A_i$, with $A_i \cap A_j = \emptyset$, $i \neq j$
(ii) $\varphi(\mathbf{x}) = \psi(\alpha_1(\mathbf{x}^{A_1}), ..., \alpha_r(\mathbf{x}^{A_r}))$.

The three modules theorem (cf. [BARL 75]).

Let there be a coherent system (C, φ). Let us assume that A_1, A_2 and A_3 are three non-empty and disjoint sets such that $A_1 \cup A_2$ and $A_2 \cup A_3$ are modular sets. Then:

(i) A_1, A_2 and A_3 are modular sets,
(ii) $A_1 \cup A_2 \cup A_3$ is modular,
(iii) the moduless $\alpha_{A_1}, \alpha_{A_2}, \alpha_{A_3}$ appear either in series or in parallel,
(iv) $A_1 \cup A_3$ is modular.

2.3 Elementary structure systems

The distinction between elementary structure systems and complex structure systems is very useful in matters of reliability. The elementary structure systems are concerned with series structure and the parallel structure or, more precisely, the structure k-over-n, which is a generalization of the preceding two structures. Any system giving a k-over-n structure through successive reductions is called a elementary structure system. In the contrary case, it is known as a complex structure system. The structural function of an elementary structure system is obtained directly using the functions of structures, as described in the following section. On the contrary, for the complex structure systems, we require the notions of minimal cut set or of minimal path.

2.3.1 *Series system*

A system is said to be in series if its functioning is subjected to the simultaneous functioning of all its components. If just one of its components is failed, then the system will have broken down.

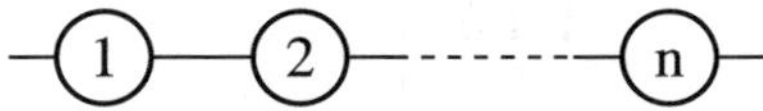

Figure 2.1 *Series system*

The structure function of a series system is given by:

$$\varphi(\mathbf{x}) = \min(x_1, ..., x_n) = \prod_{i=1}^{n} x_i.$$

The dual of a series system is a parallel system.

2.3.2 *Parallel system*

The functioning of this system is assured, if at least one of its components is in good state. The system will be failed if and only if all its components are simultaneously failed.

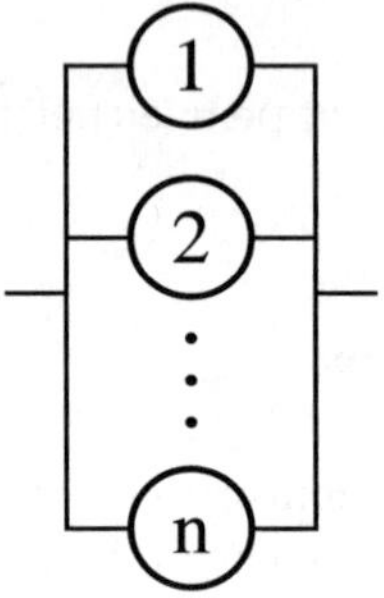

Figure 2.2 *Parallel system*

The structure function of a system is given by:

$$\varphi(\mathbf{x}) = \max(x_1, ..., x_n) = x_1 \dot{+} \cdots \dot{+} x_n = 1 - \prod_{i=1}^{n}(1 - x_i)$$

The dual of a parallel system is a series system.

2.3.3 *System k-out-of-n*

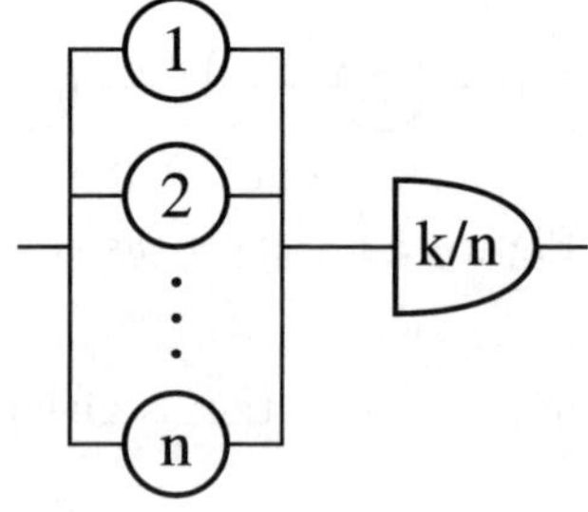

Figure 2.3 *System k-out-of-n*

The functioning of this system is assured if at least k components among n $(1 \leq k \leq n)$ are in good state. The system will be failed if $n - k + 1$ or more components are failed simultaneously.

The structural function of a system k-out-of-n:G is given by:

$$\varphi(\mathbf{x}) = \begin{cases} 1 \text{ if } \sum_{i=1}^{n} x_i \geq k \\ 0 \text{ if not.} \end{cases}$$

The series system is a system n-over-n:G, and the parallel system 1-out-of-n:G. The dual of a system k-out-of-n:G is a system $(n - k + 1)$-out-of-n:F.

2.3.4 *Parallel-series system*

This is formed by r blocks mounted in series, and each block i constitutes a parallel system of i_j components $j = 1, ..., r$.

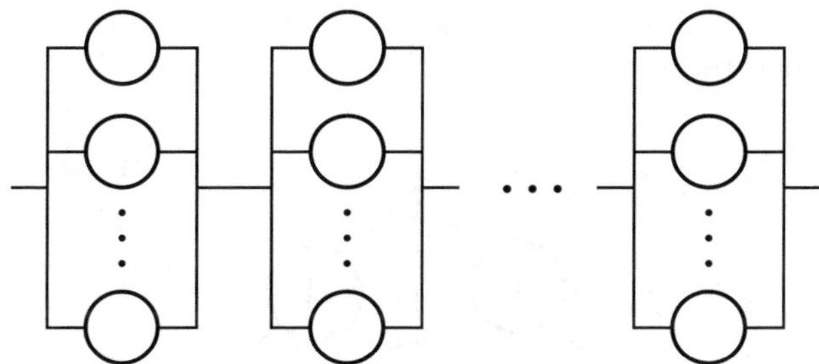

Figure 2.4 *Parallel-series system*

2.3.5 *Series-parallel system*

This is formed by r blocks mounted in parallel, and each block i constitutes a series system of i_j components $j = 1, ..., r$.

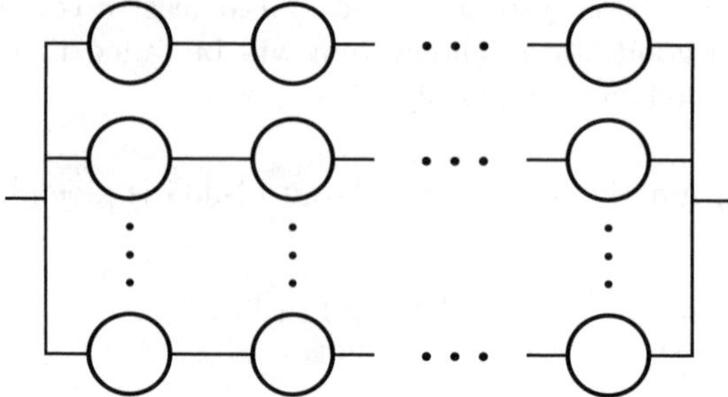

Figure 2.5 *Series-parallel system*

2.4 Systems with complex structure

A system that cannot be directly classified into the preceding cases of elementary structures is called a system with complex structure. For the study of systems with complex structure, the concepts of minimal paths and cuts (in the case of monotone structures) must be introduced. The system in Figure 2.6 is a system with complex structure. This system cannot be directly classified into modules with traditional structures. If the input/output of this system are located at the extremities of component 3, it will then involve a system with elementary structure.

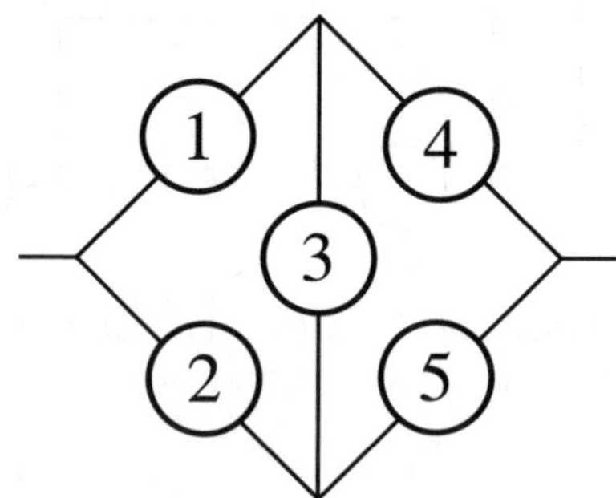

Figure 2.6 *Example of a system with complex structure*

The definitions are given as follows.

Path: a sub-set of components whose simultaneous good functioning will assure the good working of the system, which is independent of the states of the other components.

Minimal path: a path which does not contain another path.

Cut set: a sub-set of components whose simultaneous failure leads to the system failure, which is independent of the states of the other components.

Minimal cut set: a cut set that does not contain another cut set.

▷ **Example 2.2.** The set $\{1, 4, 5\}$ is a path (non-minimal), which contains the minimal path $\{1, 4\}$. The set $\{1, 2, 3\}$ is a cut set (non-minimal) and it contains the minimal cut set $\{1, 2\}$.

The minimal path sets of the system given in Figure 2.6 are:

$$C_1 = \{1, 4\}, \quad C_2 = \{2, 5\}, \quad C_3 = \{1, 3, 5\}, \quad C_4 = \{2, 3, 4\}.$$

The minimal cut sets are:

$$K_1 = \{1, 2\}, \quad K_2 = \{4, 5\}, \quad K_3 = \{1, 3, 5\}, \quad K_4 = \{2, 3, 4\}.$$

In order to study a complex system, the following hypotheses are made:

– The initial system is equivalent to the system formed by its minimal paths in parallel, where each path is represented by a series system having as components the components of the path. With reference to this hypothesis, the structural function will be given by the following relationship:

$$\varphi(\mathbf{x}) = 1 - \prod_{j=1}^{c} \Big(1 - \prod_{i \in C_j} x_i\Big),$$

where c is the number of minimal paths of the system.

– The initial system is equivalent to the system formed by its minimal cut sets in series, where each cut set is represented by a parallel system having as components the components of the cut set. With reference to this hypothesis, the structural function is given by the following relationship:

$$\varphi(\mathbf{x}) = \prod_{j=1}^{k} \Big[1 - \prod_{i \in K_j} (1 - x_i)\Big],$$

where k is the number of minimal cut sets of the system.

2.5 Probabilistic study of the systems

2.5.1 *Introduction*

Let there be a coherent system $S=(C,\varphi)$ of the order $n\geq 1$.

Let X_i, an r.v. (random variable) of Bernoulli, describe the state of the component i $(i=1,...,n)$, with values $x_i\in\{0,1\}$, and let $\mathbf{X}=(X_1,...,X_n)$ denote the vector with value $\mathbf{x}=(x_1,...,x_n)\in\{0,1\}$. We also note that $\overline{X}_i:=1-X_i$ and $\overline{\mathbf{X}}=\mathbf{1}-\mathbf{X}=(1-X_1,...,1-X_n)$.

We have: $E[X_i]=P\{X_i=1\}=:p_i$ and $E[\overline{X}_i]=P\{\overline{X}_i=1\}=P\{X_i=0\}$ $=:q_i$, $i\in C$, and $\mathbf{p}=(p_1,...,p_n)$; $\mathbf{q}=(q_1,...,q_n)$; $\mathbf{p}+\mathbf{q}=\mathbf{1}=(1,...,1)$.

The problem here consists of expressing the reliability of the system, noted as $R(\mathbf{p})$ or $R(p_1,...,p_n)$, according to the reliabilities of its components. Several methods that are exact or approximate are proposed for performing this calculation. It can be written that:

$$R(\mathbf{p})=E[\varphi(\mathbf{X})]=\sum_{\mathbf{x}}\varphi(\mathbf{x})P\{\mathbf{X}=\mathbf{x}\}=\sum_{\mathbf{x}:\varphi(\mathbf{x})=1}\prod_{i=1}^{n}q_i^{x_i}(1-q_i)^{1-x_i}.$$

This formula gives an exact method of calculation, but we will see that there are other methods that are either based or not based on the minimal sets and give quicker calculations.

As we have seen earlier, the minimal sets (minimal paths and cut sets) constitute a general method for expressing the structure function of the system and hence for calculating its reliability. In order to evaluate the reliability through the method of minimal sets, we consider two different approaches: one based on the structure function and the other based on the lifetime.

Let $\mathcal{K}=\{K_1,K_2,...,K_k\}$ be the set of minimal cut sets and $\mathcal{C}=\{C_1,C_2,...,C_c\}$ the set of minimal paths of a system (C,φ).

It is clear that:

$$R(\mathbf{p})=P[\varphi(\mathbf{X})=1]=E[\varphi(\mathbf{X})].$$

Therefore, it can be written as:

$$R(\mathbf{p})=P(C_1\cup C_2\cup\cdots\cup C_c),$$

and

$$\overline{R}(\mathbf{p})=P(K_1\cup K_2\cup\cdots\cup K_k)$$

with $\overline{R}(\mathbf{p})=1-R(\mathbf{p})$, called *unreliability* of the system.

The approach based on the lifetimes of the components and of the system, is obtained from the following equalities:

$$T = \min_{1 \leq j \leq k} \max_{i \in K_j} \{T_i\} = \max_{1 \leq j \leq c} \min_{i \in C_j} \{T_i\}$$

where T is the lifetime of the system, and T_i, $i = 1, ..., n$, the lifetime of the components.

The reliability can be obtained through different methods; the main methods will be described in the following sections.

Remark 2.2. We will use the same symbol for designating a minimal cut set, its indicating variable and the event "occurrence of the cut set".

2.5.2 *Inclusion-exclusion method*

This development, for k events, is written as:

$$P(\cup_{i=1}^{k} K_i) = \sum_{i=1}^{k} P(K_i) - \sum_{i=1}^{k-1} \sum_{j=i+1}^{k} P(K_i \cap K_j) + \cdots + (-1)^{k-1} P(\cap_{i=1}^{k} K_i). \tag{2.1}$$

The probability for the occurrence of a cut set is evaluated as follows. Let the cut set be $K_i = \{i_1, i_2, ..., i_s\}$, its probability being:

$$\begin{aligned} P(K_i) &= P(X_{i_1} = 0, X_{i_2} = 0, ..., X_{i_s} = 0) \\ &= P(X_{i_1} = 0) P(X_{i_2} = 0 | X_{i_1} = 0) \times \\ &\quad \cdots \times P(X_{i_s} = 0 | X_{i_1} = 0, ..., X_{i_{s-1}} = 0). \end{aligned}$$

In the case where the events are independent, we can write:

$$P(K_i) = P(X_{i_1} = 0) P(X_{i_2} = 0) \cdots P(X_{i_s} = 0).$$

Probability for the simultaneous occurrence of two cut sets: let the cut sets be K_i and K_j: $K_i = \{a_1, ..., a_r, i_{r+1}, ..., i_s\}$ and $K_j = \{a_1, ..., a_r, j_{r+1}, ..., j_l\}$

and hence: $K_i \cap K_j = \{a_1, ..., a_r\}$. Then:

$$\begin{aligned} P(K_i \cap K_j) &= P(K_i | K_j) P(K_j) \\ &= P(X_{i_{r+1}} = 0, ..., X_{i_s} = 0, X_{a_1} = 0, \\ &\quad ..., X_{a_r} = 0, X_{j_{r+1}} = 0, ..., X_{j_l} = 0). \end{aligned}$$

▷ **Example 2.3.** Let us consider the system of Figure 2.6. The minimal cut sets are given in section 2.4 and $\mathbf{p} = (p_1, ..., p_5)$. We have:

$$K_1 = \{1, 2\}, \quad K_2 = \{4, 5\}, \quad K_3 = \{1, 3, 5\}, \quad K_4 = \{2, 3, 4\}.$$

Consequently:

$$\begin{aligned} \overline{R}(\mathbf{p}) &= (q_1 q_2 + q_4 q_5 + q_1 q_3 q_5 + q_2 q_3 q_5) \\ &\quad - (q_1 q_2 q_4 q_5 + q_1 q_2 q_3 q_5 + q_1 q_2 q_3 q_4 + q_1 q_3 q_4 q_5 + q_2 q_3 q_4 q_5 \\ &\quad + q_1 q_2 q_3 q_5) \\ &\quad + (q_1 q_2 q_3 q_4 q_5 + q_1 q_2 q_3 q_4 q_5 + q_1 q_2 q_3 q_4 q_5 + q_1 q_2 q_3 q_4 q_5) \\ &\quad - q_1 q_2 q_3 q_4 q_5. \end{aligned}$$

2.5.3 *Disjoint products*

Let us consider the events: $E_1, E_2, ..., E_n$.

– *Disjunction of events.* We can write:

$$E_1 \cup E_2 \cup \cdots \cup E_n = E_1 \cup \overline{E}_1 E2 \cup \cdots \cup \overline{E}_1 \cdots \overline{E}_{n-1} E_n. \tag{2.2}$$

This relationship enables us to express the disjunction of n initial events, which are not necessarily disjunctive, into a disjunction of n disjunctive events. Consequently, we can write:

$$P\{E_1 \cup E_2 \cup \cdots \cup E_n\} = P\{E_1\} + P\{\overline{E}_1 E_2\} + \cdots + P\{\overline{E}_1 \cdots \overline{E}_{n-1} E_n\}. \tag{2.3}$$

– *Conjunction of events.* Let us consider the event: $E = E_1 E_2 \cdots E_n$ (we are omitting $\cap$). We can write:

$$\overline{E} = \overline{E}_1 \cup E_1 \overline{E}_2 \cup \cdots \cup E_1 \cdots \cdots E_{n-1} \overline{E}_n.$$

For the indicator variables of the events mentioned above, $x_1, x_2, ..., x_n$ we have:

$$x_1 \dot{+} x_2 \dot{+} ... \dot{+} x_n = x_1 + \overline{x}_1 x_2 + ... + \overline{x}_1 ... \overline{x}_{n-1} x_n.$$

If $x = x_1 x_2 .... x_n$, we have:

$$\overline{x} = \overline{x}_1 + x_1 \overline{x}_2 + ... + x_1 x_2 ... x_{n-1} \overline{x}_n. \qquad (2.4)$$

▷ **Example 2.4.** Let us consider the system in Figure 2.6. We saw that its minimal paths are:

$$C_1 = \{1,4\}, \quad C_2 = \{2,5\}, \quad C_3 = \{1,3,5\}, \quad C_4 = \{2,3,4\}.$$

Let us note c_i, the indicator variable of the minimal path C_i, for $i = 1,2,3,4$. According to (2.4), we can write:

$$c_1 = x_1 x_4, \quad \text{which implies} \quad \overline{c}_1 = \overline{x}_1 + x_1 \overline{x}_4;$$

$$c_2 = x_2 c_5, \quad \text{which implies} \quad \overline{c}_2 = \overline{x}_2 + x_2 \overline{x}_5;$$

and

$$c_3 = x_1 x_3 x_5, \quad \text{which implies} \quad \overline{c}_3 = \overline{x}_1 + x_1 \overline{x}_3 + x_1 x_3 \overline{x}_5.$$

According to (2.2), we have:

$$C = C_1 \cup C_2 \cup C_3 \cup C_4 = C_1 \cup \overline{C}_1 C_2 \cup \overline{C}_1 \overline{C}_2 C_3 \cup \overline{C}_1 \overline{C}_2 \overline{C}_3 C_4,$$

But then, if c is the indicator variable of the event C, we have:

$$\begin{aligned} c = {} & x_1 x_4 + (\overline{x}_1 + x_1 \overline{x}_4) x_2 x_5 + (\overline{x}_1 + x_1 \overline{x}_4)(\overline{x}_2 + x_2 \overline{x}_5) x_1 x_3 x_5 \\ & + (\overline{x}_1 + x_1 \overline{x}_4)(\overline{x}_2 + x_2 \overline{x}_5)(\overline{x}_1 + x_1 \overline{x}_3 + x_1 x_3 \overline{x}_5) x_2 x_3 x_4, \end{aligned}$$

and on simplifying, we obtain:

$$c = x_1 x_4 + \overline{x}_1 x_2 x_5 + x_1 x_2 \overline{x}_4 x_5 + x_1 \overline{x}_2 x_3 \overline{x}_4 x_5 + x_1 x_2 x_3 \overline{x}_4 \overline{x}_5 + \overline{x}_1 x_2 x_3 x_4 \overline{x}_5,$$

From which:

$$R(\mathbf{p}) = p_1 p_4 + q_1 p_2 p_5 + p_1 p_2 q_4 p_5 + p_1 q_2 p_3 q_4 p_5 + q_1 p_2 p_3 p_4 q_5 + p_1 p_2 p_3 q_4 q_5.$$

Remark 2.3. The formulae (2.1) and (2.2) can be very easily proved through recursion.

2.5.4 *Factorization*

The technique of factorization is based on the following relationship:

$$\varphi(\mathbf{x}) = x_i\varphi(1_i, \mathbf{x}) + (1 - x_i)\varphi(0_i, \mathbf{x}). \tag{2.5}$$

This relationship is known in other works on this subject (given in the bibliography) as the formula or development or theorem of Shannon (or the formula of factorization). We will continue the decomposition until such time as we arrive at the elementary structures.

▷ **Example 2.5.** The system is as given in Figure 2.6. The above relationship can be written with respect to the variable x_3 as follows:

$$\varphi(\mathbf{x}) = x_3\varphi(1_3, \mathbf{x}) + (1 - x_3)\varphi(0_3, \mathbf{x}).$$

The functions $\varphi(1_3, \mathbf{x})$ and $\varphi(0_3, \mathbf{x})$ represent two systems with elementary structures (cf. Figure 2.7).

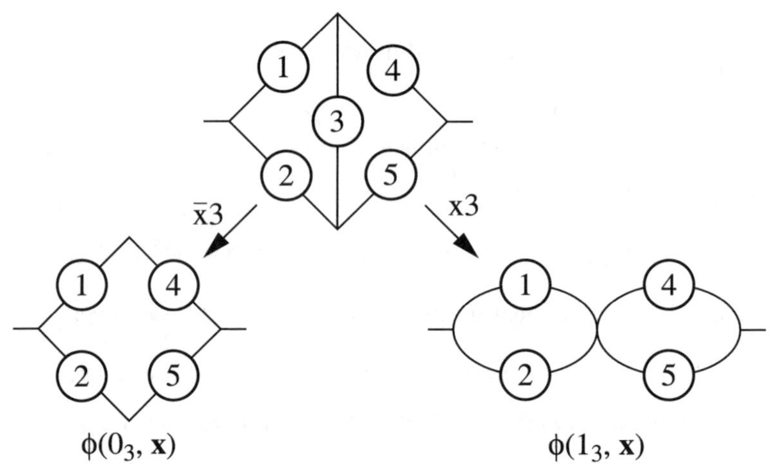

Figure 2.7 *Decomposition of a system by factorization*

2.5.5 *Reliability bounds*

The exact calculation for the reliability of the systems is very costly (NP-hard problem); that is why the calculation of the bounds, instead of for exact value, is suitable. We have four types of bounds: the bounds of inclusion–exclusion, the minimal sets bounds, the “min–max” bounds and the trivial bounds. Other bounds can be obtained by improving the above-cited bounds or through the construction of mixed bounds or by a modularization of the system. It has to be noted that there is a relationship of order among those four types of bounds. The best bound depends on the problem.

Inclusion-exclusion bounds

The inclusion-exclusion bounds are closely linked to the inclusion-exclusion development (cf. section 2.5.2). They are:

$$\sum_{i=1}^{k} P(K_i) - \sum_{i=1}^{k-1} \sum_{j=i+1}^{k} P(K_i \cap K_j) \leq R(\mathbf{p}) \leq \sum_{i=1}^{k} P(K_i). \tag{2.6}$$

For reliable systems, the lower bound is very close to the exact value. For less reliable systems, an improvement in the bounds can be obtained by calculating more terms in the inclusion-exclusion development. An even number of terms of this development yields a lower bound, while an odd number of terms gives an upper bound. Theoretically, these bounds are also valid for the minimal paths of the system. Nevertheless, this does not lead to any significant bounds, because for the reliable systems they lie outside the interval [0, 1]. It must be mentioned that we simply require the minimal cut sets for these bounds.

Minimal sets bounds

A lower bound is obtained through minimal cut sets, and an upper bound through minimal paths, as follows:

$$\prod_{j=1}^{k} \left(1 - \prod_{i \in K_j} (1 - p_i)\right) \leq R(\mathbf{p}) \leq 1 - \prod_{j=1}^{c} \left(1 - \prod_{i \in C_j} p_i\right). \tag{2.7}$$

When the minimal cut sets are 2-by-2 disjoint sets, then the upper bound coincides with the exact value. The same assertion is valid regarding the minimal paths and the lower bound.

"Min–max" bounds

The complement, with respect to unity, of the probability of the most probable cut set is an upper bound, whereas the complement, with respect to unity, of the probability of the least probable path is a lower bound:

$$\max_{1 \leq j \leq c} \left[1 - \prod_{i \in C_j} p_i\right] \leq R(\mathbf{p}) \leq \min_{1 \leq j \leq k} \left[1 - \prod_{i \in K_j} (1 - p_i)\right]. \tag{2.8}$$

Trivial bounds

The reliability of a coherent system is higher than that of a series system and lower than that of a parallel system. This assertion is expressed as follows:

$$\prod_{i=1}^{n} p_i \leq R(\mathbf{p}) \leq 1 - \prod_{i=1}^{n}(1 - p_i). \tag{2.9}$$

▷ **Example 2.6.** Let us consider the system given in Figure 2.6.

– inclusion-exclusion bounds:

$$\begin{aligned}(q_1q_2 + q_4q_5 + q_1q_3q_5 + q_2q_3q_5) - (q_1q_2q_4q_5 + q_1q_2q_3q_5 + q_1q_2q_3q_4 \\ + q_1q_3q_4q_5 + q_2q_3q_4q_5 + q_1q_2q_3q_5) \leq R(\mathbf{p}) \\ \leq q_1q_2 + q_4q_5 + q_1q_3q_5 + q_2q_3q_5.\end{aligned}$$

– minimal sets bounds:

$$\begin{aligned}&(1 - q_1q_2)(1 - q_4q_5)(1 - q_1q_3q_5)(1 - q_2q_3q_4) \leq R(\mathbf{p}) \\ &\leq 1 - (1 - p_1p_4)(1 - p_2p_5)(1 - p_1p_3p_5)(1 - p_2p_4p_5).\end{aligned}$$

– min–max bounds:

$$\begin{aligned}\max\{(p_1p_4), (p_2p_5), (p_1p_3p_5), (p_2p_4p_5)\} \leq R(\mathbf{p}) \\ \leq \min\{(1 - q_1q_2), (1 - q_4q_5), (1 - q_1q_3q_5), (1 - q_2q_3q_4)\}.\end{aligned}$$

– trivial bounds:

$$p_1p_2p_3p_4p_5 \leq R(\mathbf{p}) \leq 1 - q_1q_2q_3q_4q_5.$$

Chapter 3

Construction of Fault Trees

3.1 Basic ideas and definitions

We first give out the basic notions constituting the vocabulary of fault trees before moving on to the stages of construction and treatment of fault trees (FT).

Graphic symbol	Name	Meaning
	OR	The output is generated if at least one of the inputs exists
	AND	The output is generated if all the inputs exist

Table 3.1 *Fundamental operators*

Graphic symbol	Name	Meaning
	Exclusive OR	The output is generated if one and only one input exists
	Priority or sequential IF	The output is generated if all the inputs exist, with an order of appearance
C	IF	The output is generated if the input exists and if the state C is verified
k/n ... A1...An	K-out-of-n combination	The output is generated if k-out-of-n inputs exist ($1 \leq k \leq n$)
...	Matrix	The output is generated for certain combinations of inputs
Δt	Delay	The output is generated with a delay Δt over the input that should be present during Δt
	No	The output is generated when the input is not produced

Table 3.2 *Special operators*

Graphic symbol	Meaning
	Rectangle Top or intermediate event
	Circle Elementary basic event
	Rhombus Non-elementary basic event
	Double rhombus Event that is considered to be basic in this step and will be analyzed later
	House Event considered as being normal

Table 3.3 *Events*

Graphic symbol	Meaning
	Identical transfer The part of the tree that should follow is not indicated, as it is identical to the part tagged by the last symbol
	Similar transfer The part of the tree that should follow is not indicated, as it is similar to the part tagged by the last symbol
	Identification of the transfer Marks an identical or similar subtree that is not otherwise resumed

Table 3.4 *Transfer triangles*

3.1.1 *Graphic symbols*

The fault tree is represented by means of three types of graphic symbols:

– The logic gates or operators: fundamental operators, as given in Table 3.1 and special operators, as indicated in Table 3.2. This list is not exhaustive, as other operators are not included here.

– The events, given in Table 3.3.

– The transfer triangles, as given in Table 3.4.

The transfer triangles are utilized to make the representation of the fault tree more compact, by avoiding repetitions.

An operator is said to be *primary* when all its inputs are basic events.

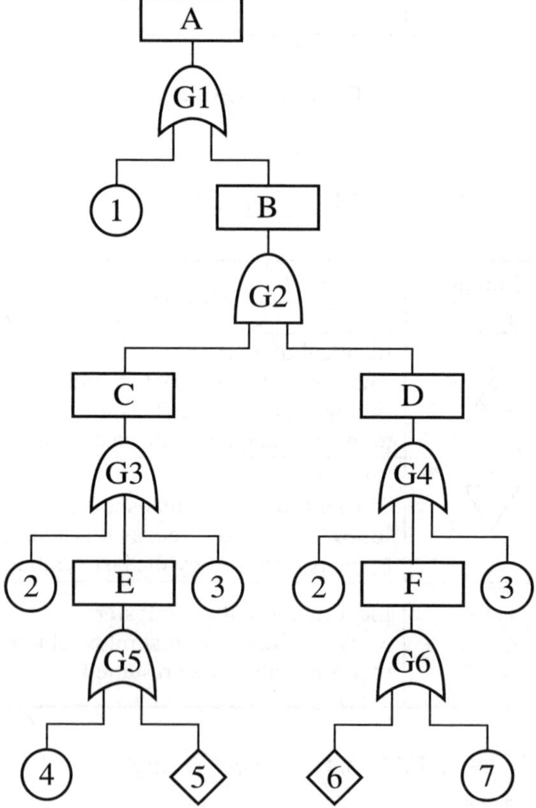

Figure 3.1 *Fault tree*

▷ **Example 3.1.** Figure 3.1 gives an example of an FT. This FT is composed of:

– Top event: A.

– Intermediate events: B, C, D, E and F.

– Basic events: $1, 2, 3, 4, 5, 6$ and 7, where the events 5 and 6 are not elementary.

– Events 2 and 3, which are repeated events.

– Operators "OR": G_1, G_3, G_4, G_5 and G_6. Operator "AND": G_2.

Among these operators, the operators G_5 and G_6 are primary, because all their inputs are basic events.

An FT can have unique representation through the fundamental operators ("AND" and "OR"). This representation is called the restrained form. We designate the set of basic events of the fault tree by $\mathcal{E}$, that is, $\mathcal{E} = \{e_1, ..., e_N\}$. The numbers assigned to the basic events on the fault tree correspond to the indices of the events, that is, the number i corresponds to the event $e_i \in \mathcal{E}$. We associate an indicator variable with every event. If an FT, under its restrained form, contains two complementary events, we say that it contains biform variables; in the opposite case, we say that it contains solely the monoform variables.

The sub-FT corresponding to an intermediate event is the biggest FT having this intermediate event as its top event.

The domain of a sub-FT corresponding to the intermediate event X is the subset of $\mathcal{E}$ containing the basic events of the subtree X, denoted as $\mathcal{D}(X)$.

▷ **Example 3.2.** All the variables associated with the FT in Figure 3.1 are monoforms, because none of these events have their respective complementary event.

The sub-FT corresponding to the intermediate event D is the FT containing the operators G_4, G_6 and the basic events, which are inputs of these operators. The domain of this sub-FT is: $\mathcal{D}(D) = \{2, 3, 6, 7\}$.

3.1.2 *Use of the operators*

We give examples for the use of operators regarding the FT-c and temporal operators. As for the operators concerning the FT-nc, their equivalent structures are given in Chapter 7.

– *OR operator*: this operator describes the failure of a series system (Figure 3.2).

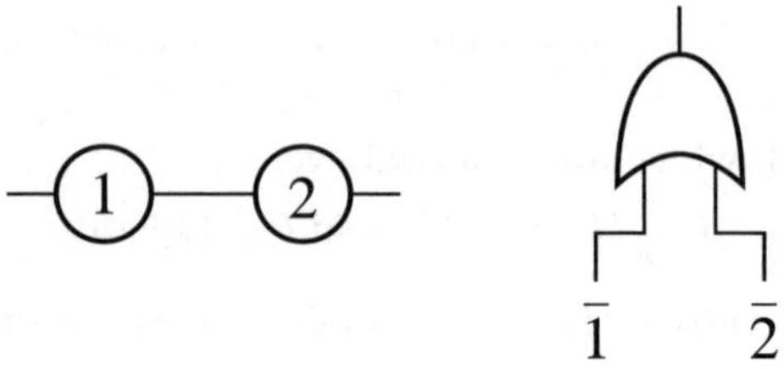

Figure 3.2 *Series system and the OR operator*

– *AND operator*: this operator describes the failure of a parallel system (Figure 3.3).

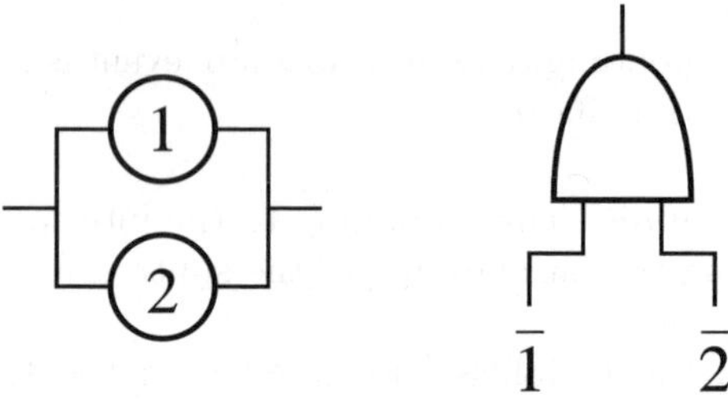

Figure 3.3 *Parallel system of active redundancy and the AND operator*

– *Priority AND operator*: this describes the failure of a parallel system with passive redundancy (Figure 3.4). Unlike in the case of the preceding redundancy, called the active redundancy, this passive redundancy involves the functioning of component 1, while component 2 is on standby. When the component 1 breaks down, component 2 starts functioning instantaneously. Now, the failure of the system necessitates at first the failure of component 1 (it is agreed that component 2 cannot break down, when it is on standby) and then the failure of component 2 after having started functioning.

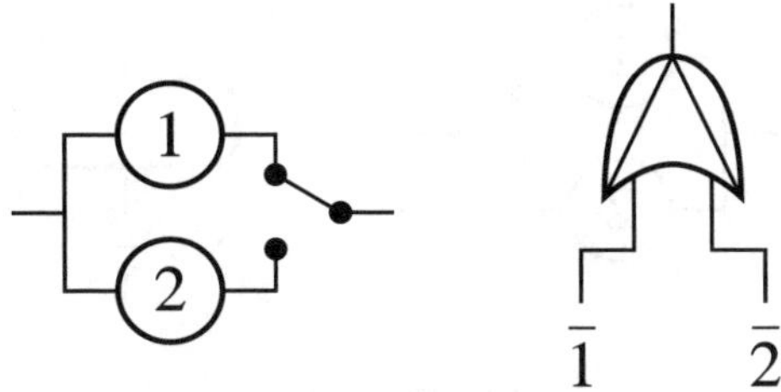

Figure 3.4 *Parallel system with passive redundancy and the priority AND operator*

– *IF operator*: from the logical viewpoint, that is, the Boolean equation, it is equivalent to an ordinary AND operator. Nevertheless, it cannot be confused with an ordinary AND operator as it plays quite a different role in the construction of a fault tree. It is used for briefly presenting an event without analyzing it through its cause events, either because this analysis is not possible, at least at the required level of precision, or because it is too time-consuming. To assist understanding of the function of this operator, we can say that the input event A designates the state of the entity, and the event C is an external event with respect to the space of this entity's states, and its occurrence is uniquely possible when the entity is situated within the states designated by the event A. The probability p of the state C is a conditional probability $p = P(C|A)$, where A is the input event of the operator. This operator is very much used in safety studies. For example, the explosion of an industrial machine can result the death of the user if a splinter happens to hit him. Here we make use of an IF operator because we cannot analyze the trajectories of the machine's fragments and the position of the user to be able to predict whether he will be hit by the flying projectiles. The top event will be "death of the user", the input event will be "explosion of the machine" and the state will be "a splinter hits the user".

– *DELAY operator*: let us consider an electric system made up of a main source of supply (1) and a secondary source of supply (2), as passive redundancy, whose role consists of making up for the failures in the main source. The lifetime of (1) is random, while that of (2) is fixed and is equal to τ ($\tau > 0$). Thus, when (1) breaks down, (2) comes into play. The system breaks down only when the time taken for repairing (1) is greater than τ. In the opposite case, the system does not break down. The failure of the system is represented by the DELAY operator in Figure 3.5.

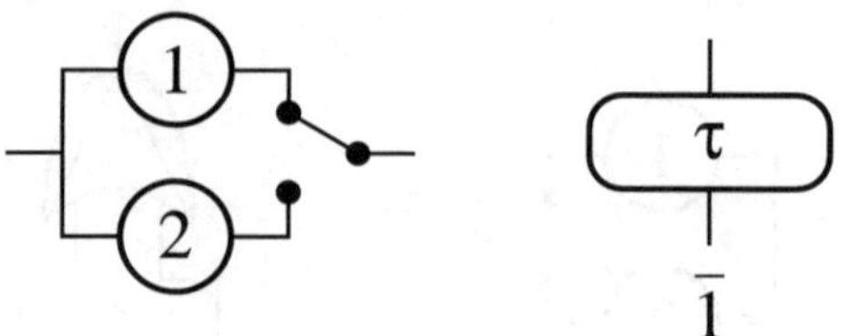

Figure 3.5 *DELAY operator*

3.2 Formal definition and graphs

The representation of an FT by a standard graph offers many advantages with regard to its construction and its treatment. There are numerous algorithms dealing with the graphs, which can be applied almost directly to the FTs.

From the graphical viewpoint, we give the formal definition as follows:

An FT is a 1*-graph, quasi strongly connected, without loop and without circuit* (Figure 3.6).

Let $G = (X, U)$ be this graph, where X is the set of vertices and U the set of edges.

Each event is represented by a vertex of G; the intermediate events are identified by the corresponding operators. The arcs link the intermediate events with the inputs of the corresponding operators.

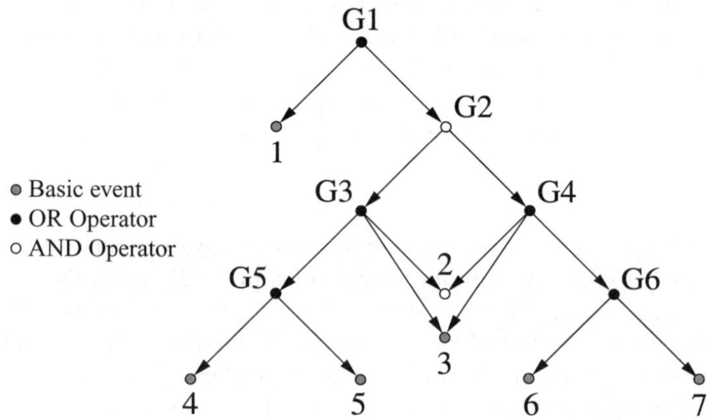

Figure 3.6 *Fault tree graph*

Type	Representation	Comments
Loop	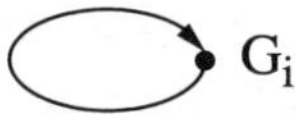 G_i	When an operator is its own input
Non-connected operator	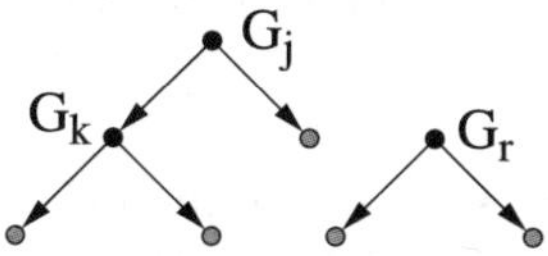 G_j, G_k, G_r	When an operator (≠top) is not an input to another operator
Pending operator	 G_i, G_j	An operator without input
Circuit	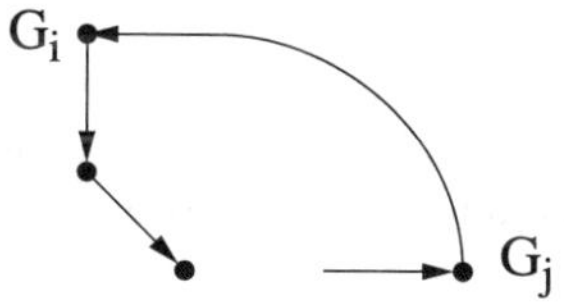 G_i, G_j	An operator has for input one of its predecessors
Multi-graph	 G_i, j	An operator has for input the same event more than once

Table 3.5 *Configurations that are not admitted in an FT graph*

The vertex of G, corresponding to the top event of the FT, is called the root of an FT graph. From the definition of the FT itself, we deduce the configurations that are not allowed in an FT graph. They are given in Table 3.5.

3.3 Stages of construction

The construction is a very vast task that requires a deep knowledge of the system that is being studied. This implies the "horizontal" knowledge, of the complex systems, we see here an overlapping of the most diverse disciplines (physics, chemistry, electronics, automatic control engineering, computer science, etc.) and "vertical" knowledge, because the fidelity of the representation of the "undesirable" event, defined at the level of the system through FT, depends on the precise definition of the logical links existing among the different components of this system and its failure modes.

Thus, the construction of an FT should be the fruit of collaboration among the different specialists, who intervene in the realization of the system, from the designer to the operator entrusted with the running of the system.

The construction starts off by defining the undesirable event, still called the top event. This event is resolved into "intermediate events". The intermediate events are in turn developed until such time as any new resolving becomes impossible or also judged as being useless; this last possibility then implies the knowledge of quantitative data such as the probabilities of final events, called the basic events.

Subsequently, we are going to present a general approach for the systematic construction of the FTs. This consists of three phases: preliminary analysis, specifications and the construction.

3.3.1 *Preliminary analysis*

(a) Decomposition of the system: this involves a physical decomposition of the system. The criteria that are generally used are as follows:

– Criteria of technology: for example, a microprocessor that controls an electric circuit will be separately taken into account.

– Criteria of maintenance: for example, when a part of the system is systematically replaced in the wake of a failure.

– Criteria of data on the study made: for example, when for a part of the system, the data are adequate or a particular study has already been carried out.

It is also possible to envisage other criteria.

(b) Identification of the components: this involves the identification of all the devices that are represented at the last resolving level of the system; within the framework of our analysis, we call them "components".

(c) Definition of failure modes of the components: for each component, the possible failure modes should be defined; that is to say, the different manners manifested by the failure.

(d) Reconstitution of the system through the components: it is necessary to reconstitute the system into functional mode by climbing back the levels of decomposition.

Remark 3.1. More details of all these information are provided in the FMEA (Fault Modes and Effects Analysis) tables (see for example [VIL 88]).

3.3.2 *Specifications*

(a) Phases: we refer to the different working modes of a system as phases. Almost all the systems have many working modes. For example, in the case of an aircraft in flight, we have at least three phases: takeoff, flight at altitude and landing.

(b) Boundary conditions: these are concerned with the interactions of the system with its environment.

(c) Specific hypotheses: these are concerned with the conventions on the system itself.

(d) Initial conditions: these refer to the hypotheses in connection with the commencement of the phase under study.

3.3.3 *Construction*

(a) Defining the undesirable event: the undesirable event (top event) to be studied must be defined without ambiguity and in a coherent manner with the preceding specifications.

(b) Resolution of the events: this deals with the resolving of the undesirable event into its immediate cause events and the resolving of the latter into their own cause-events, etc.

(c) End of construction: construction is over when all the non-resolved cause events are failure modes of the components or of the environment.

During the construction, it should be remembered that the FT is a purely deductive tool, hence considerable caution is called for when proceeding from the general to the particular. The general scheme of proceeding is based on three types of failures (in a more general manner, we speak of "faults"): primary fault, secondary fault and control defect. For example, at the component level, the primary fault is a failure mode and involves an elementary basic event (circle), the secondary fault a non-elementary basic event (rhombus) and the control fault an intermediate event.

The secondary fault is characterized by an abnormal functioning of the entity. The operating factors comprise amplitude, frequency and duration. The variables of the environment of the system are of thermal, mechanical, electrical, chemical, magnetic and electromagnetic nature.

When resolving an intermediate event, it is imperative that all of its causes be defined before their analysis is taken up. Each operator input should have its own text, and there shall be no two operators with direct connection between them.

3.4 Example of construction

With a view to capture the approach that is elaborated above in a better manner, we construct a small FT.

Let us consider the hydraulic system given below (Figure 3.7):

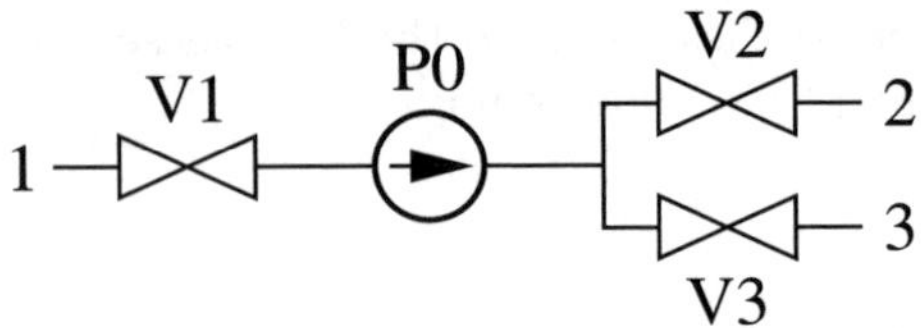

Figure 3.7 *Hydraulic system*

This is designed for transporting water from the point (1) to the areas of its consumptions (2) and (3). It contains the valves V_1, V_2 and V_3, the centrifugal pump P_0, and the adjacent pipes with hydraulic components.

3.4.1 *Preliminary analysis*

Decomposition of the system:
Level 1:

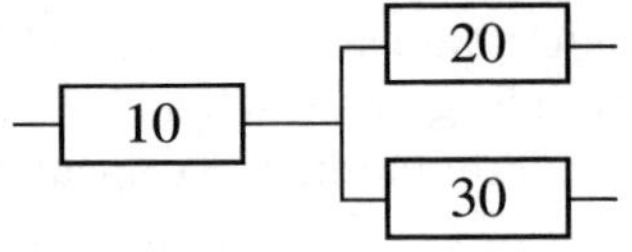

Level 2:

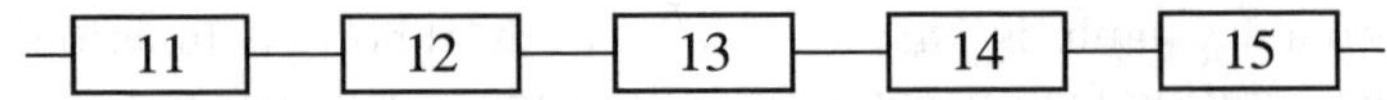

Block 10 contains the valve, pump and the adjacent pipes.

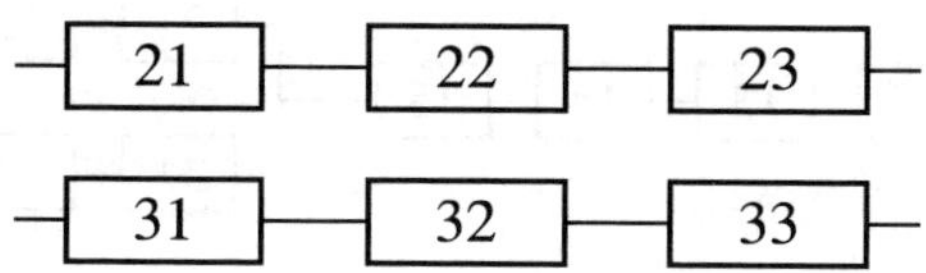

Each of the blocks 20 and 30 contains one valve and two adjacent pipes.

The system is represented as follows:

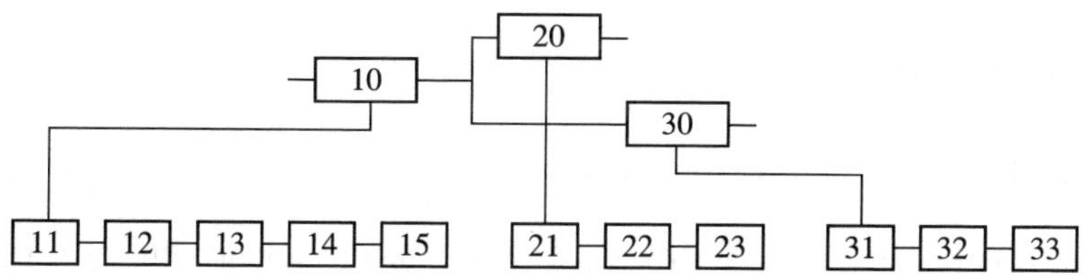

Identification of the components:

The components are: 11, 12, 13, 14, 15; 21, 22, 23; 31, 32, 33.

Definition of the failure modes of the components:

Valve (12, 22, 32)
- MD1: opened blocked
- MD2: closed blocked
- MD3: ill-timed closure
- MD4: ill-timed opening

Pump (P0)
- MD1: out of service

Pipe
- MD1: clogging
- MD2: leakage
- MD3: crack

Reconstitution of the system through its components:

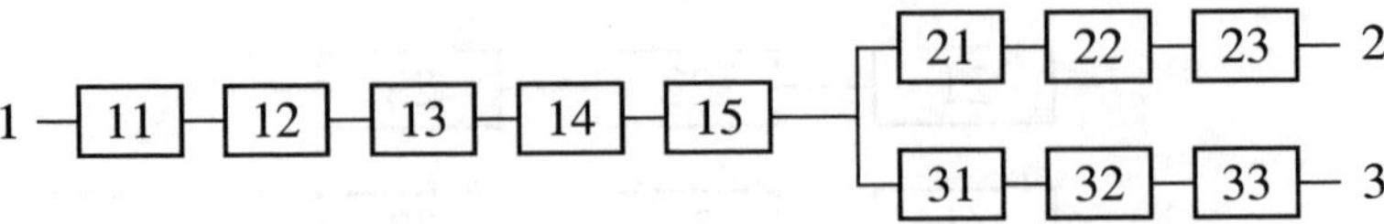

3.4.2 *Specifications*

– Phase:

In normal functioning

– Boundary conditions:

– Availability of water at the point 1,

No other interaction with regard to the environment will be considered (i.e. cracking of a pipe due to an external cause, etc.)

– Specific hypotheses:

The pipes will not be taken into account in this study.

– Initial conditions:

The system functions normally at the beginning of the working phase with a flow of 100%.

3.4.3 *Construction*

Definition of the undesirable event:

"Total stopping of the flow"

(i.e. the event "not starting" would not adhere to the preceding specifications).

The construction of the tree is presented in Figure 3.8.

3.5 Automatic construction

In the studies on operational safety, where the FT intervenes, its construction is the most important task, as it conditions the other stages, that is, the qualitative analysis and the evaluation. It is also the most time-consuming and the most difficult to perform. These reasons have, from the beginning, led the analysts to conceive of systematic techniques of construction with a view to automate this stage.

Two sets of techniques have been developed in this regard: one based on the decision tables and the other on the graphs, depending on the type of representation of the system being studied.

The first consists of modeling the components of the system through the decision tables that translates the relationships among the inputs, the internal states (failure modes and the different operational configurations) and the outputs of each component. This method, called Cat (Computer-Aided Trees [APO 76], [SAL 80]), is adapted for the discrete systems, the easily discretizable ones, such as the electric systems.

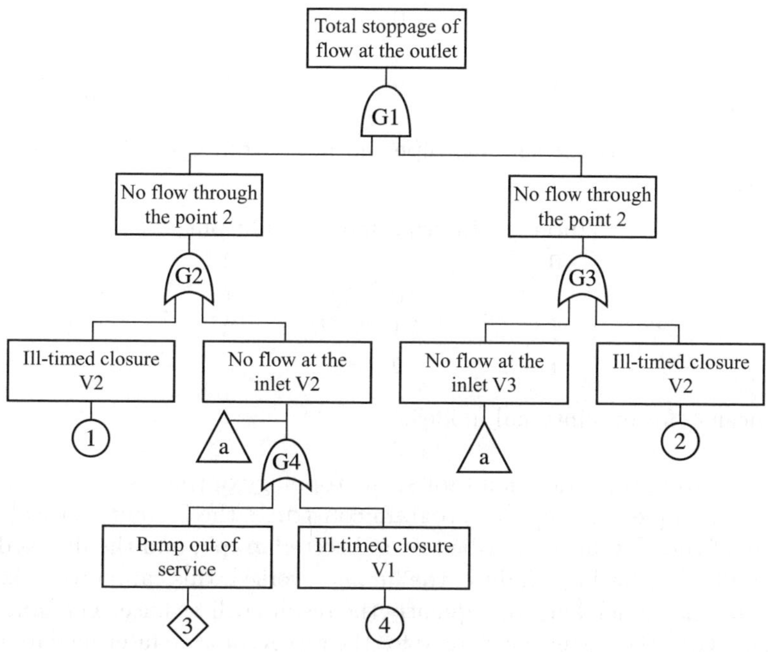

Figure 3.8 *Fault tree of the hydraulic system*

▷ **Example 3.3.** In the case of the preceding hydraulic system, a model of pipe through a decision table is:

Input and output	Failure mode	Normal mode
0: zero flow	1: clogging	0: normal functioning
1: normal flow	2: leakage	

The decision table is as shown below:

Perfect pipe:

Inlet	Outlet
0	0
1	1

Real pipe (with the failure modes):

Input	Internal mode	Output
0	0	0
0	1	0
0	2	0
1	0	1
1	1	0
1	2	0

The decision table for the real pipe can be written under the reduced form as follows:

Input	Internal mode	Output
0	X	0
1	0	1
1	1	0
1	2	0

X means "for any internal mode".

Starting from these decision tables, the topology of the system and the definition of the top event, the Cat program constructs the FT automatically. The topology of the system is described by a directed graph, and the flux is defined once for all. If, following a failure, the flux is reversed, this cannot be taken into account by the model, and consequently the results will be false. The same problem arises when the vertex considered for the top event is an intermediate vertex. These disadvantages are discussed in detail in [CAR 86], where a novel method Cafts (Computer-aided Fault-Tree Synthesis) is proposed for offsetting the problems encountered in Cat. It replaces the decision tables by the "logical" equations, which renders the model much more flexible. Unlike in the case of Cat, Cafts is not entirely automatized but is interactive with the user.

A second set of techniques that are entirely based on modeling through the directed graphs is proposed in [LAP 77] and [LAP 79]. The vertices of the graph represent the variables of the system and certain types of failures (i.e. flow, temperature, pressure, etc.) and the arcs represent the relationships among the variables. If a change in a variable V_1 brings about a change in another variable V_2, then these two variables (vertices of the graph) are connected by a weighted edge having a value h, i.e.

$$V_1 \overset{\text{h}}{\longrightarrow} V_2$$

If $\Delta V_1 \cong \Delta V_2$, then $h = 1$, if $\Delta V_1 \gg \Delta V_2$, then the edge $(h = 0)$ is cancelled, and if $\Delta V_1 \ll \Delta V_2$, then $h = 10$. If the variations are in opposite directions, then h will be negative.

Starting from this representation and from the definition of the top event that takes place relatively at a vertex of the graph, the software Fts [LAP 76] automatically constructs the FT.

The advantages of an automatic construction are evident:

– It allows a systematic analysis, which guarantees a very high degree of exhaustivity of the construction.

– Once the system is modeled by a graph, a decision table, or similar parameters, the construction of more FTs dealing with the same system requires practically no additional cost.

– It favors a considerable reduction in certain types of studies carried out mainly through FT, such as the studies of safety.

The disadvantages are as follows:

– At present, there are no general models for dealing with the majority types of existing systems.

– Large simplifications are done *a priori* for the systems of continuous state spaces and the dynamic systems.

– There are risks of systematic errors.

In [AND 80], there is a noteworthy study that focuses on the difficulties existing during the automatic construction of the FTs.

Chapter 4

Minimal Sets

4.1 Introduction

When the construction of an FT is over, it can be analyzed from an algebraic point of view (this analysis is also called logical or even qualitative analysis); that is, we have to determine the structure function (or indicator function) of the FT concerning the top event, or, generally, any intermediate event. The determination of the structure function involves the study of minimal sets, that is, minimal cut sets and minimal path sets.

It has to be noted that the structure function is the indicator function of the top event of the FT; that is, it takes up the value 1 for the occurrence of the top event, and the value 0 for its non-occurrence. In a binary system, the top event of the FT translates the failure of the system, and consequently the structure function takes up the value 1 when the system has broken down. This is contrary to the convention in Chapter 2.

The construction and reduction of the structure function of the FT is the most significant in this analysis. The structure function of an FT will be dual in comparison with the FT given in Chapter 2, with an inversion of the convention adopted, i.e.:

$$x_i = \begin{cases} 1, \text{ if the event i occurs (failure)} \\ 0, \text{ otherwise (functioning)} \end{cases}$$

For all values of $i \in \mathcal{E}$:

Same convention for the structure function:

$$\varphi(\mathbf{x}) = \varphi(x_1, \ldots, x_n) = \begin{cases} 1, \text{ if the top event occurs} \\ 0, \text{ otherwise} \end{cases}$$

We admit that the FT contains:

– uniquely the operators AND and OR, and

– the monoform variables.

In other words, the FT is coherent.

A basic event is said to be *relevant* or *essential* if its indicator variable appears in the minimal form of $\varphi(\mathbf{x})$. In the opposite case, it is said to be irrelevant or inessential.

We give here the definitions of the minimal sets adapted in the context of the FTs:

Path: a subset of events, whose simultaneous non-existence involves the non-occurrence of the top event, and which is independent of the occurrence or non-occurrence of the other events of the FT.

Minimal path: a path that does not contain another path.

Cut set: a subset of events, whose simultaneous existence involves the occurrence of the top event, and which is independent of the occurrence or non-occurrence of the other events of the FT.

Minimal cut set: a cut set that does not contain another cut set.

4.2 Methods of study

4.2.1 *Direct methods*

We can directly construct on the FT, its structure function and give its minimal sets by means of the latter. The approach adopted here consists of:

– the construction of $\varphi(\mathbf{x})$,

– its development,

– its reduction for obtaining the minimal form.

Each term of the minimal form represents a minimal cut set of the FT.

The construction of $\varphi(\mathbf{x})$ is based on the following relationships:

Let the operator be $E(A_1, \ldots, A_s)$, where E is the output event, and $A_1, \ldots, A_s$ are the input events of the operator.

– for an OR operator, we have: $E = A_1 \cup \ldots \cup A_s$,

– for an AND operator, we have: $E = A_1 \cap \ldots \cap A_s$.

For the indicator variables we have:

– OR operator: $y_E = x_1 \dot{+} \cdots \dot{+} x_s$,

– AND operator: $y_E = x_1 \cdots x_s$.

where y_E is the indicator variable of the intermediate event E.

The minimal form is obtained by the following reductions:

$$x^n = x,$$
$$nx = x, \quad (n \in \mathbb{N}^*),$$
$$x \dot{+} xy = x.$$

▷ **Example 4.1.** Let us consider the FT CH (Figure 4.1).

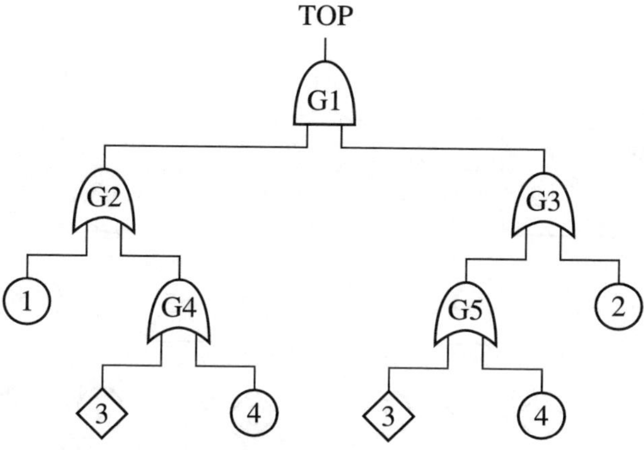

Figure 4.1 *FT CH*

The indices for the operators of this FT correspond to the indices of the indicator variables y_i of the intermediate events.

Construction of $\varphi(\mathbf{x})$:

$$\varphi(\mathbf{x}) = y_1 = y_2 y_3 = (x_1 \dot{+} y_4)(x_2 \dot{+} y_5) = (x_1 \dot{+} x_3 \dot{+} x_4)(x_2 \dot{+} x_3 \dot{+} x_4).$$

Development:

$$\varphi(\mathbf{x}) = x_1 x_2 \dot{+} x_1 x_3 \dot{+} x_1 x_4 \dot{+} x_3 x_2 \dot{+} x_3 x_3 \dot{+} x_3 x_4 \dot{+} x_4 x_2 \dot{+} x_4 x_3 \dot{+} x_4 x_4.$$

Reduction:

$$\varphi(\mathbf{x}) = x_1 x_2 \dot{+} x_3 \dot{+} x_4.$$

The minimal cut sets are:

$$K_1 = \{3\}, \quad K_2 = \{4\}, \quad K_3 = \{1, 2\}.$$

The minimal paths are obtained in the same way as minimal cut sets, but by proceeding with the dual FT.

The *dual* fault tree is obtained from the fault tree, by replacing the AND gates with the OR gates, the OR gates with the AND gates, and the events with their complementary events.

▷ **Example 4.2.** Let us consider the dual FT of the FT CH in Figure 4.2.

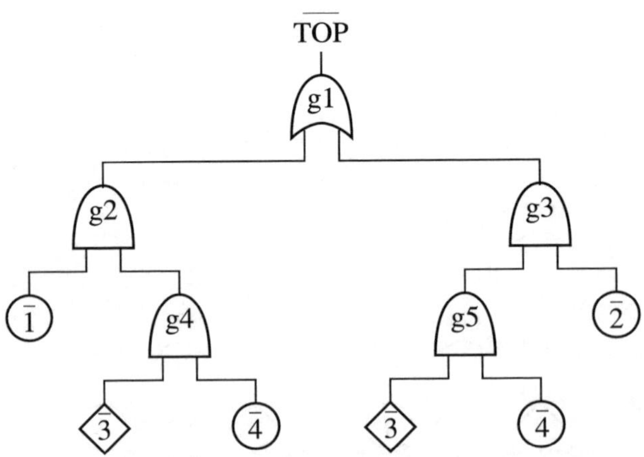

Figure 4.2 *Dual FT of the FT CH*

We have:

$$\overline{\varphi}(\mathbf{x}) = \overline{y}_1 = \overline{y}_2 \dot{+} \overline{y}_3 = \overline{x}_1\overline{x}_3\overline{x}_4 \dot{+} \overline{x}_2\overline{x}_3\overline{x}_4.$$

The above form is already minimal. Hence the minimal paths are:

$$C_1 = \{1,3,4\} \quad \text{and} \quad C_2 = \{2,3,4\}$$

Within the framework of the coherent FTs, we have two types of methods. These are descending methods, proceeding from the top operator and descending the FT by decomposing the operators, and ascending methods, which proceed in the opposite direction.

4.2.2 *Descending methods*

We present the algorithm *Mocus*, which is the oldest and the most used. This algorithm was proposed by Fussell and Vesely [FUS 72]. It consists of initializing a matrix B through the top operator and in resolving it into its inputs. When the input itself happens to be an operator, it will also be resolved in the succeeding stage and so on, until such time that all the elements of the matrix B are basic events. Each row of the matrix B obtained in the last stage represents a cut set. The reduction of these cut sets provides us the list of the minimal cut sets in the FT.

The decomposition of the operators in the matrix B is done as follows.

Algorithm 4.1. Mocus

0. Initialize the first element of a matrix with the top event operator.
1. The operator $G_i(A_1, \ldots, A_s)$ occupying the place (i, j) of the matrix B_k is to be resolved at the stage k.
2. If it is an AND operator, we will replace it with its inputs in the row. The first input takes the place of the operator, and the subsequent inputs the places $(i, j+1), (i, j+2), \ldots, (i, j+s-1)$.
3. If it is an OR operator, we will replace it with its inputs in the column. The first input takes the place of the operator, and the subsequent inputs the places $(i+1, j), (i+2, j), \ldots, (i+s, j)$.
   In addition, each element $b_{i,m}$, $m = 1, \ldots, s$, $m \neq i$, will be repeated (Figure 4.3). The blocks B_1 and B_2 remain unaltered.
4. If there is another operator in B, then continue as per 1.

End.

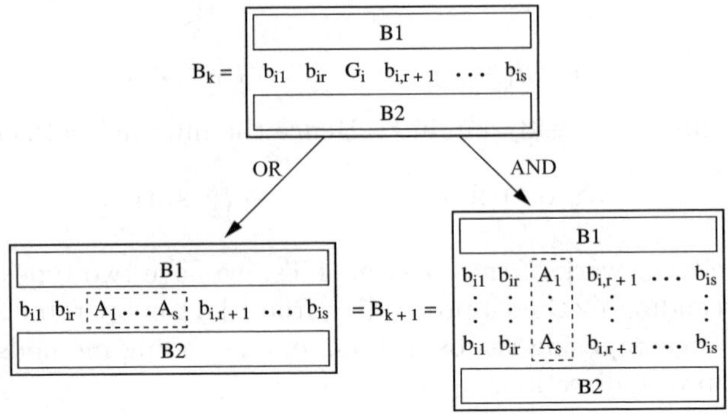

Figure 4.3 *Resolving the operators in the MOCUS algorithm*

▷ **Example 4.3.** Application of the Mocus algorithm to the FT CH. Cut sets:

$$B_1 = [\underline{G_1}], \quad B_2 = [\underline{G_2}, G_3], \quad B_3 = \begin{bmatrix} 1 & \underline{G_3} \\ G_4 & G_3 \end{bmatrix},$$

$$B_4 = \begin{bmatrix} 1 & 2 \\ 1 & \underline{G_5} \\ G_4 & G_3 \end{bmatrix}, \quad B_5 = \begin{bmatrix} 1 & 2 \\ 1 & 3 \\ 1 & 4 \\ \underline{G_4} & G_3 \end{bmatrix}, \quad B_6 = \begin{bmatrix} 1 & 2 \\ 1 & 3 \\ 1 & 4 \\ 3 & \underline{G_3} \\ 4 & G_3 \end{bmatrix},$$

$$B_7 = \begin{bmatrix} 1 & 2 \\ 1 & 3 \\ 1 & 4 \\ 3 & 2 \\ 3 & \underline{G_5} \\ 4 & \underline{G_3} \end{bmatrix}, \quad B_8 = \begin{bmatrix} 1 & 2 \\ 1 & 3 \\ 1 & 4 \\ 3 & 2 \\ 3 & 3 \\ 3 & 4 \\ 4 & 2 \\ 4 & \underline{G_5} \end{bmatrix}, \quad B_9 = \begin{bmatrix} 1 & 2 \\ 1 & 3 \\ 1 & 4 \\ 3 & 2 \\ 3 & 3 \\ 3 & 4 \\ 4 & 2 \\ 4 & 3 \\ 4 & 4 \end{bmatrix}.$$

The operator that is analyzed at each stage is underlined.

When the matrix B_9, is obtained, the algorithm is completed as all its elements are basic events. Each line of this matrix corresponds to a cut set.

We find here the development de φ given in the preceding section. After reducing these cut sets, we obtain the minimal cut sets given in the same section.

The paths will be obtained by applying the same algorithm on the dual FT.

4.2.3 *Ascending methods*

The ascending methods proceed from the basic events by climbing up the fault tree up to the top event. At each stage, the obtained cut sets (or the paths) are reduced. Thus, we also obtain the minimal cut sets of all the intermediate events.

First, we proceed from the primary operators by replacing the intermediate events corresponding to these operators with the Boolean expression of its inputs. Then, we proceed again in the same manner from the operators that have become primary: obtaining and reducing from their Boolean expressions. The procedure is over once the Boolean expression that is reduced from the top event is obtained.

▷ **Example 4.4.**

Stage 1: the primary operators are G_4 and G_5. We replace them with their inputs.

$$y_4 = x_3 \dot{+} x_4 \quad \text{and} \quad y_5 = x_3 \dot{+} x_4.$$

Stage 2: the operators G_2 and G_3 have become primary. Now,

$$y_2 = x_1 \dot{+} x_3 \dot{+} x_4, \quad \text{there is no reduction,}$$

$$y_3 = x_2 \dot{+} x_3 \dot{+} x_4, \quad \text{there is no reduction.}$$

Stage 3: the operator G_1 has become primary. Now,

$$\begin{aligned} y_1 &= (x_1 \dot{+} x_3 \dot{+} x_4)(x_2 \dot{+} x_3 \dot{+} x_4) \\ &= x_1x_2 \dot{+} x_1x_3 \dot{+} x_1x_4 \dot{+} x_2x_3 \dot{+} x_3x_3 \dot{+} x_3x_4 \dot{+} x_2x_4 \dot{+} x_3x_4 \dot{+} x_4x_4. \end{aligned}$$

And after reduction,

$$\varphi(\mathbf{x}) = y_1 = x_1x_2 \dot{+} x_3 \dot{+} x_4.$$

The Miscup algorithm, proposed by Chatterjee [CHA 74], is the first of this type. The development of big FTs, having AND operators near the top, yield quite a considerable number of Boolean terms. In order to avoid this, Nakashima & Hatori [NAK 79] proposed a very interesting algorithm, called Anchek.

4.3 Reduction

The reduction of the obtained cut sets and paths is the most time-consuming task. For a given number n of cut sets, $n!/[2(n-2)!]$ operations of comparison at the maximum have to be performed. This number increases when the number of cut sets is considerable.

The Mocus algorithm, applied to an FT not containing any repeated event, directly yields the minimal cut sets without reduction.

We have shown [LIM 86] that the cut sets not containing any repeated event are minimal. The reduction is then solely limited to the cut sets containing repeated events.

A reduction algorithm is as follows.

Algorithm 4.2. LZ

1. Obtain the cut sets $\mathcal{K}'$ (for example: Mocus);
   If the FT does not contain repeated event, then: $\mathcal{K} = \mathcal{K}'$; end.
   If no, continue the stage 2;
2. Partition $\mathcal{K}'$ into two subsets $\mathcal{K}'_1$ and $\mathcal{K}'_2$
   such that the former comprises of all the cut sets
   containing at least one repeat event;
3. Reduce the cut sets in $\mathcal{K}'_1$, let $\mathcal{K}_1$ be the reduced set;
4. $\mathcal{K} = \mathcal{K}_1 \cup \mathcal{K}'_2$;

End.

$\mathcal{K}$ is the set of minimal cut sets of the FT.

▷ **Example 4.5.** Consider the FT in Figure 4.4.

The Mocus algorithm generates the following 9 cut sets:

$$\{1\}, \{2\}, \{3\}, \{6\}, \{8\}, \{4,6\}, \{4,7\}, \{5,7\}, \{5,6\}.$$

For reducing these cut sets, 36 comparisons should be made. By restricting oneself solely to the cut sets containing the repeated event 6, that is, the cut sets $\{6\}, \{4,6\}$ and $\{5,6\}$, (the other cut sets already being minimal), we have to carry out 3 comparisons. The minimal cut sets are:

$$\{1\}, \{2\}, \{3\}, \{6\}, \{8\}, \{4,7\}, \{5,7\}.$$

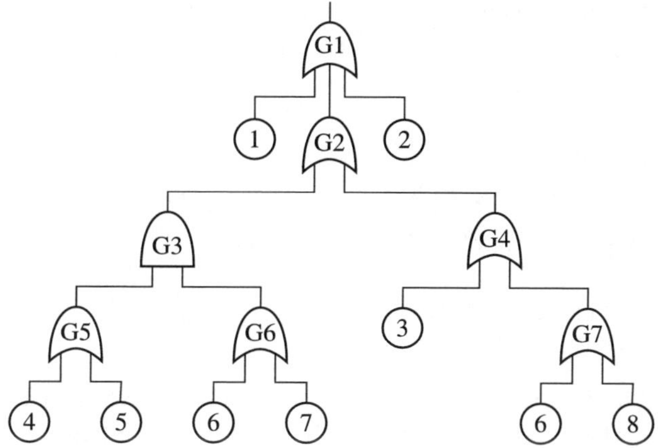

Figure 4.4 *FT for the qualitative analysis*

4.4 Other algorithms for searching the cut sets

Many other algorithms have been developed for obtaining the minimal cut sets of FTs. The majority of these algorithms have tried to improve upon the Mocus algorithm by taking into consideration the repeated events. Among the most important of such algorithms, we have the algorithm of Bengiamin *et al.* [BEN 76] and the algorithm of Fatram [RAS 78]. In the following sections, we will be describing Fatram with a view to illustrate the inclusion of the repeated events when the cut sets are constructed.

Algorithm 4.3. Fatram

1. Reduce FT by deleting all the primary OR operators on replacing them with basic events.
2. Find all the minimal cut sets (i.e. through the application of the Mocus algorithm).
3. Treat each repeated event in the inputs of the OR operators. For each repeated event:
    a. The repeated event replaces all the OR operators that are not yet resolved and for which this event is an input;
    b. These new sets are added in the collection;
    c. This event is eliminated from the OR operators, in which it appears;
    d. Eliminate all the non-minimal cut sets;
4. Resolve the remaining OR operators; all the sets are minimal cut sets.

End.

▷ **Example 4.6.** Application of the Fatram algorithm on the FT of Figure 4.4.
Stages 1 and 2: $\mathcal{K}_0 = \{\{1\},\{2\},\{3\},\{G_5,G_6\},\{G_7\}\}$.
Stage 3: repeated event 6

a. $\{\{6\},\{G_5,6\}\}$;
b. $\mathcal{K}_1 = \mathcal{K}_0 \cup \{\{6\},\{G_5,6\}\} = \{\{1\},\{2\},\{3\},\{G_5,G_6\},\{G_7\},\{6\},\{G_5,6\}\}$
c and d. $\mathcal{K}_2 = \{\{1\},\{2\},\{3\},\{G_5,G_6\},\{G_7\},\{6\}\}$.

Stage 4: $\mathcal{K} = \{\{1\},\{2\},\{3\},\{6\},\{8\},\{4,7\},\{5,7\}\}$.

Remark 4.1. The Fatram algorithm, as mentioned above, is a more compact, new version of the original algorithm [RAS 78]. As well as the algorithm of Bengiamin *et al.*, it can bring about an improvement over the Mocus algorithm, only in the case of the FT having primary OR operators. In the opposite case, they do not offer any improvement in terms of execution time.

4.5 Inversion of minimal cut sets

The problem of inversion consists of obtaining the minimal paths (or the minimal cut sets) from the minimal cut sets (or from the minimal paths), respectively. This operation is useful from the practical viewpoint, as it allows us to obtain a type of minimal set from the preceding type, without again having to take recourse to the FT.

The existing algorithms are based on the De Morgan equations:

$$\overline{xy} = \overline{x} \dotplus \overline{y},$$

$$\overline{x \dotplus y} = \overline{x} \cdot \overline{y}.$$

▷ **Example 4.7.** Let $K_1 = \{1,2\}$, $K_2 = \{3\}$ and $K_3 = \{4\}$ be the minimal cut sets of the FT CH (Figure 4.1). The structure function φ is written as:

$$\varphi(\mathbf{x}) = x_1 x_2 \dotplus x_3 \dotplus x_4.$$

On applying the equalities of De Morgan, we get:

$$\overline{\varphi}(\mathbf{x}) = (\overline{x}_1 \dotplus \overline{x}_2)\overline{x}_3\overline{x}_4.$$

By development (here, there is no reduction), we get:

$$\overline{\varphi}(\mathbf{x}) = \overline{x}_1\overline{x}_3\overline{x}_4 \dotplus \overline{x}_2\overline{x}_3\overline{x}_4.$$

This gives rise to the minimal paths of the FT:

$$C_1 = \{1,3,4\}, \quad C_2 = \{2,3,4\}.$$

The algorithm SW [SHI 85] is based on an approach proposed initially in [LOC 78].

Another algorithm, called "Inmin", is proposed for the inversion of the minimal sets. For an FT $T(\mathcal{E})$, with $\mathcal{E} = \{e_1,\ldots,e_N\}$, $\mathcal{K}$ is the set of its minimal cut sets (given), and $\mathcal{C}$ the set of its minimal paths (to be calculated). Let $\mathcal{S}$ be the set of all the subsets of $\mathcal{E}$, sequenced as per the lexicographic order. In addition, let: $\mathcal{K}^* = \{K = \mathcal{E} \setminus H \mid H \in \mathcal{K}\}$.

Algorithm 4.4. Inmin ([HEI 83])

1. Calculate $\mathcal{K}^*$;
2. Put $\mathcal{I} = \emptyset$ et $\mathcal{C} = \emptyset$;
3. If $(\mathcal{I} \notin \mathcal{J}, \forall \mathcal{J} \in \mathcal{K})$ and $(\mathcal{L} \notin \mathcal{I}, \forall \mathcal{L} \in \mathcal{C})$, then $\mathcal{C} := \mathcal{C} \cup \mathcal{I}$;
4. If $\mathcal{I} \neq \mathcal{E}$, then replace $\mathcal{I}$ by its successor in $\mathcal{S}$, and continue the stage 3;

End.

If we permute $\mathcal{K}$ and $\mathcal{C}$ in the preceding algorithm, we obtain the minimal cut sets from the minimal paths.

▷ **Example 4.8.** Let $\mathcal{K} = \{\{1,2\},\{1,3\}\}$. The Inmin algorithm gives:

1. $\mathcal{K}^* = \{\{3\},\{2\}\}$;

2. $\mathcal{I} = \emptyset$ and $\mathcal{C} = \emptyset$;

3. for $\mathcal{I} = \{1\}$ there is no set $\mathcal{J}$ of $\mathcal{K}^*$ containing $\mathcal{I}$:

$\mathcal{C} = \{\{1\}\}$;

for $\mathcal{I} = \{2,3\}$ there is no set $\mathcal{L}$ of $\mathcal{C}$ contained in $\mathcal{I}$:

$\mathcal{C} = \{\{1\},\{2,3\}\}$.

Here, $\mathcal{C}$ is the final set of minimal paths, because for all the other elements $\mathcal{I}$ of $\mathcal{S}$, $\mathcal{C}$ does not change.

4.6 Complexity of the search for minimal cut sets

The time for the execution of an algorithm depends on the size of the problem, machine, programming language, etc. To use the abstraction of the machine, in general the time of execution is measured as the number of elementary operations that have to be carried out. In fact, the time of execution is described according to the problem size, perhaps the data size, the number of towns for the problem of the salesman, the order of a square matrix for the algorithm for calculating its inverse, the number of vertices in a graph, etc. The increase in the time of calculation according to the data for a type of problem can be characterized as being polynomial, exponential or more. This increase is a measure of the algorithmic complexity.

An exponential increase or more renders the large-sized problems untreatable. The cost of an algorithm in terms of the number of elementary operations is defined by the maximum of necessary operations over all the data of the same size.

In general, for the study of the algorithmic complexity, we are not interested in the exact number of the elementary operations but in the order of the size only described by the Landau notation. In particular, with regard to $O(g)$, where g describes the size of the problem's data, a function f is said to be in $O(g)$ in the neighborhood V of x, if $|f(x)| \leq C|g(x)|$, for all x in V.

Three classes of problems are of particular interest from the viewpoint of their algorithmic complexity: the class P, the class NP and the class NP-complete. The class P includes all the problems resolvable by a determinist Turing machine in polynomial time; that is to say, the time of execution of a problem of this class, whose data size is N, is of the order $O(N^a)$, where a is a positive constant.

The algorithms concerning the problems of the class P are said to be efficient. The NP class includes the problems resolvable by a non-determinist Turing machine in polynomial time. In a more precise manner, NP is the class of decision problems, from which one can verify that an instance is correct in a polynomial time. The NP-complete class is a subclass of the NP class. Any problem of the NP class can be transformed in a polynomial time into a problem of the class NP-complete. This class includes the problems that are very difficult to resolve, and no algorithm of polynomial order has been found for any of these problems.

The possibility of finding the polynomial algorithms for the NP-hard problem is an open problem. When a polynomial algorithm for one of the problems is obtained, the polynomial algorithms for all the other problems are also obtained. For example, the saleman problem is a NP-complete problem. As for the search for the minimal cut sets of the FT, this problem is of the NP class [ROS 75], [ODE 95].

Chapter 5

Probabilistic Assessment

5.1 The problem of assessment

The probabilistic assessment of an FT consists of calculating the probability of a top event starting from the probabilities of the basic events. This can be done directly when the FT does not possess any repeated event. This is carried out with a simple approach, which consists of climbing back up the FT by starting from its primary operators up to the top event.

When the FT possesses repeated events, the calculation as mentioned above is no longer applicable (it yields over-evaluated results). For exact calculation in this case, we should pass through minimal sets of the FT and then use one of the methods presented in Chapter 2; that is, the inclusion-exclusion development or the disjoint products or factorization. In Chapter 9, we will give a more recent method based on binary decision diagrams.

Because of the high reliability of the majority of the systems, the inclusion-exclusion development is the most commonly used, as the second term only yields highly satisfactory results.

Two families of problems, from the viewpoint of the probabilistic assessment, are usually raised.

Problem 1: non-repairable systems.

For these systems, we have: $R(t) = A(t) = 1 - P_S(t)$, for all $t \geq 0$. The second equality is valid in the case of binary systems. In the opposite case, we have to replace $P_S(t)$ with $\sum_{i=1}^{r} P_{S_i}(t)$ where $S_i, i = 1, ..., r$ indicates the r failure modes of the system.

The data are $\{\lambda_i(t)\}_{i\in A}$, $\{q_j\}_{j\in B}$, $\{\gamma_k\}_{k\in C}$, $\{A, B, C\}$ is a partition of $\mathcal{E}$, where $t \rightarrow \lambda_i(t)$, $t \geq 0$, $i \in A$ is the function of the occurrence rate of the basic event i of the FT (the failure rate of a component of the system),

q_j, $i \in B$ is the probability of a basic event over a time interval $[0, \theta]$, with θ being the duration of the mission,

γ_k, $k \in C$ is the probability of a basic event, expressing the stress failure or the probability of the condition in an IF operator.

When we have mixed data, a special attention has to be paid to the homogenity of the assessment. The calculation has to be carried out for the time θ. On the other hand, if $A = \mathcal{E}$, then the calculation can be done for all $t \geq 0$.

The result gives the probability of the top event to occur over $[0, t]$, $P_S(t)$

Problem 2: repairable systems.

For the repairable systems, we have: $R(t) \leq A(t) = 1 - P_S(t)$. It has to be noted that in the case of these systems, we cannot assess the reliability by means of the FT, thus constituting a considerable limitation of the FTs in comparison with the random processes.

Nevertheless, we can calculate an approximate value of the reliability by defining a pseudo-rate of the occurrence of basic events (cf. section 5.3.3).

The data are $\{\lambda_i(t), \mu_i(t)\}_{i\in A}$, $\{q_j\}_{j\in B}$, $\{\gamma_k\}_{k\in C}$, $\{A, B, C\}$ is a partition of $\mathcal{E}$, where $t \rightarrow \mu_i(t)$, $t \geq 0$, $i \in A$, is the function of the rate of disappearance (of repair) of the basic event i (of the component).

The result is the probability for the existence of the top event at time $t \geq 0$, i.e. $P_S(t)$.

5.2 Direct methods

When the FT does not contain any repeated event, we can obtain the probability of the top event through direct calculation over the FT, without having to go in for minimal sets or for other techniques.

For this, it is sufficient to start from the primary operators and climb up the FT by calculating the probabilities of the intermediate events, whose respective operators have known or have already calculated probability inputs. The procedure continues until the top event is reached.

The calculation at the level of each operator is done with the help of the following formulae.

5.2.1 *AND operator*

Inputs A and B, output E; equation: $E = A \cap B$.

– A and B dependent:

$$P(E) = P(A)P(B|A) = P(B)P(A|B).$$

– A and B independent:

$$P(E) = P(A)P(B).$$

– A and B mutually exclusive:

$$P(E) = 0.$$

– A and B dependent and $A \subset B$:

$$P(E) = P(A).$$

When the input data are λ, $\tau = 1/\mu$, then the stationary λ and τ of E are given by the following relationships:

$$\lambda_E = \frac{\lambda_A \lambda_B (\tau_A + \tau_B)}{1 + \lambda_A \tau_B + \lambda_B \tau_B} \quad \text{and} \quad \tau_E = \left(\frac{1}{\tau_A} + \frac{1}{\tau_B}\right)^{-1}$$

For the non-repairable case, we have:

$$\lambda_E = \min\{\lambda_A, \lambda_B\}$$

The probabilities $P(A)$ and $P(B)$ are linked with (λ_A, τ_A) and (λ_B, τ_B) through the relationships given in Chapter 1, section 1.2.4.

5.2.2 *OR operator*

Inputs A and B, output E; equation: $E = A \cup B$.

– A and B dependent:

$$P(E) = P(A) + P(B) - P(A \cap B) = P(A) + P(B) - P(A)P(B|A).$$

– A and B independent:

$$P(E) = P(A) + P(B) - P(A)P(B).$$

– A and B mutually exclusive:

$$P(E) = P(A) + P(B).$$

– A and B dependent and $A \subset B$:

$$P(E) = P(B).$$

When the input data are $(\lambda, \tau = 1/\mu)$, the stationary λ and τ of E are given by the following relationships:

$$\lambda_E = \lambda_A + \lambda_B \text{ (\textit{for the repairable and non-repairable case})}$$

and

$$\tau_E = \frac{\lambda_A \tau_A + \lambda_B \tau_B + \lambda_B \tau_A \tau_B}{\lambda_A + \lambda_B}.$$

The formulae given above, concerning AND as well as OR operators, can be easily extended to the operators having more than two inputs [LIE 75].

5.2.3 *Exclusive OR operator*

Inputs A and B, output E; equation: $E = (\overline{A} \cap B) \cup (A \cap \overline{B})$.

– A and B dependent:

$$P(E) = P(A) + P(B) - 2P(A \cap B).$$

– A and B independent:

$$P(E) = P(A) + P(B) - 2P(A)P(B).$$

– A and B mutually exclusive:

$$P(E) = P(A) + P(B).$$

– A and B dependent and $A \subset B$:

$$P(E) = P(B) - P(A).$$

– λ_E, μ_E stationary:

$$\lambda_E \cong \lambda_A + \lambda_B \quad \text{and} \quad \mu_E \cong \frac{\lambda_A + \lambda_B}{\frac{\lambda_A}{\mu_A} + \frac{\lambda_B}{\mu_B}}.$$

5.2.4 *k-out-of-n operator*

Inputs $A_1, A_2, ..., A_n$, output E. The output is generated, when at least k inputs among n $(1 \leq k \leq n)$ are generated. If $P(A_1) = P(A_2) = ... = P(A_n) = q$, then:

$$P(E) = \sum_{i=k}^{n} C_n^i q^i (1-q)^{n-i}.$$

5.2.5 *Priority-AND operator*

Inputs A and B, output E. The occurrence of the event E is obtained if we have the occurrence of event A and then the occurrence of event B; that is, the existence of the two events A and B are not sufficient for causing the occurrence of the output event E [FUSS 76].

– A and B independent:

$$P(E) = \frac{\lambda_B}{\lambda_A + \lambda_B} - \exp(-\lambda_A t) + \frac{\lambda_A}{\lambda_A + \lambda_B} \exp[-(\lambda_A + \lambda_B)t].$$

If $\lambda_i t << 1$, for $i = A, B$, we can write:

$$P(E) \cong \frac{1}{2}\lambda_A \lambda_B t^2 - 2\lambda_A t.$$

The rate of stationary occurrence λ_E and the rate of stationary disappearance μ_E are given by:

$$\lambda_E \cong \lambda_A \lambda_B / \mu_A \quad \text{and} \quad \mu_E \cong \mu_A + \mu_B.$$

5.2.6 *IF operator*

Let there be an IF operator having as input the event A, then the condition C of probability γ and as output the event E. If A and C are independent, then:

$$P(E) = \gamma \times P(A),$$

and

$$\lambda_E = \gamma \times \lambda_A.$$

5.3 Methods of minimal sets

Once the minimal sets are obtained, the problem is the same for any model: block diagram of reliability, FT, etc.

Thus, we will be using the formulation given in section 2.5; that is, $K=\{K_1,K_2,...,K_k\}$ the set of minimal cut sets, and $C=\{C_1,C_2,...,C_c\}$ the set of the minimal paths of the FT.

Let us note by $P_S = Q(\mathbf{q})$ the probability of the top event. With $\mathbf{q} = (\mathrm{q}_1,...,\mathrm{q}_n)$ the probabilities of the basic events of the FT, that is: $P(X_i=1)=q_i$ and $P(X_i=0)=p_i$, $i=1,...,n$, $(p_i+q_i=1)$, and $p=(p_1,...,p_n)$ the probabilities of the supplementary basic events.

5.3.1 *Inclusion-exclusion development*

We can write:

$$P_S = P\{K_1 \cup K_2 \cup ... \cup K_k\}$$

and

$$P_S = 1 - P\{C_1 \cup C_2 \cup ... \cup C_c\}.$$

▷ **Example 5.1.** Evaluation of the FT CH (see section 4.2) from the minimal sets.

– Evaluation from the minimal cut sets:

The minimal cut sets are:

$$K_1=\{3\}, \quad K_2=\{4\}, \quad K_3=\{1,2\},$$

now

$$\begin{aligned}
P_S &= P\{K_1\cup K_2\cup K_3\} = P\{K_1\}+P\{K_2\}+P\{K_3\} \\
&\quad -(P\{K_1\cap K_2\}+P\{K_1\cap K_3\}+P\{K_2\cap K_3\})+P\{K_1\cap K_2\cap K_3\} \\
&= P\{X_3=1\}+P\{X_4=1\}+P\{X_1=1,X_2=1\}-(P\{X_3=1,X_4=1\} \\
&\quad +P\{X_3=1,X_1=1,X_2=1\}+P\{X_4=1,X_1=1,X_2=1\}) \\
&\quad +P\{X_3=1,X_4=1,X_1=1,X_2=1\}
\end{aligned}$$

$$
\begin{aligned}
&= P\{X_3=1\}+P\{X_4=1\}+P\{X_1=1\}P\{X_2=1\}\\
&\quad -(P\{X_3=1\}P\{X_4=1\}+P\{X_3=1\}P\{X_1=1\}P\{X_2=1\}\\
&\quad +P\{X_4=1\}P\{X_1=1\}P\{X_2=1\})\\
&\quad +P\{X_3=1\}P\{X_4=1\}P\{X_1=1\}P\{X_2=1\}\\
&= q_3+q_4+q_1q_2-q_3q_4-q_3q_1q_2-q_4q_1q_2+q_3q_4q_1q_2.
\end{aligned}
$$

– Evaluation from the minimal paths:

The minimal paths are:

$$C_1=\{1,3,4\},\quad C_2=\{2,3,4\},$$

now

$$
\begin{aligned}
P_S &= 1-P\{C_1\cup C_2\}\\
&= 1-(P\{C_1\}+P\{C_2\}-P\{C_1\cap C_2\})\\
&= 1-(P\{X_1=0,X_3=0,X_4=0\}+P\{X_2=0,X_3=0,X_4=0\}\\
&\qquad -P\{X_1=0,X_2=0,X_3=0,X_4=0\})\\
&= 1-(P\{X_1=0\}P\{X_3=0\}P\{X_4=0\}+P\{X_2=0\}P\{X_3=0\}P\{X_4=0\}\\
&\qquad -P\{X_1=0\}P\{X_2=0\}P\{X_3=0\}P\{X_4=0\})\\
&= 1-(p_1p_3p_4+p_2p_3p_4-p_1p_2p_3p_4).
\end{aligned}
$$

5.3.2 *Disjoint products*

As we have already seen in Chapter 2, this method consists of starting from the minimal sets and in transforming the structure function of the FT into the sum of disjoint products. The most efficient method is that developed by Abraham [ABR 79], and its principle, for the minimal cut sets, is as follows.

Let:

$$E=K_1\cup K_2\cup\ldots\cup K_n=K_1\cup\overline{K}_1K_2\cup\ldots\cup\overline{K}_1\ldots\overline{K}_{n-1}K_n.$$

Algorithm 5.1. Abraham's algorithm

1. Put $L_0 = K_1$ and $M_0 = 1$;
2. Make for $i = 1, ..., n-1$;
   $M_i = M_{i-1}\overline{K}_i$
   $L_i = M_i K_{i+1}$
   End;
3. $E = L_0 \cup L_1 \cup ... \cup L_n$;

End.

In the same example, we have, by means of the minimal cut sets:

$L_0 = 3;\ M_0 = 1;$

$i = 1$: $M_1 = M_0\overline{K}_1 = \overline{3}$

$L_1 = M_1 K_2 = \overline{3}\,4$

$i = 2$: $M_2 = M_1\overline{K}_2 = \overline{3}\,\overline{4}$

$L_2 = M_2 K_3 = \overline{3}\,\overline{4}\,1\,2$

$E = L_0 \cup L_1 \cup L_2 = 3 \cup \overline{3}\,4 \cup \overline{3}\,\overline{4}\,1\,2.$

The detailed algorithm is given in [ABR 79].

5.3.3 *Kitt method*

Contrary to the stationary probabilistic assessments (constant probabilities for the basic events) the *Kitt* (Kinetic tree theory) method consists of a coupling of the FTs and the stochastic processes (see [VES 70]). Each basic event is described by an underlying stochastic process, and its probability is given depending on time. Proceeding from here, and by means of the minimal sets, the probability of the top event is evaluated in terms of time. Other magnitudes are also calculated, such as the average number of occurrences of an event over a time interval.

Kitt modelization is as follows: for the basic events, we define the following two quantities:

$\widehat{\lambda}(t)\Delta t + o(\Delta t)$: the probability of the basic event occurring in the time interval $(t, t + \Delta t]$, knowing that it does not exist at the time t.

$\widehat{\mu}(t)\Delta t + o(\Delta t)$: the probability of the basic event disappearing in the time interval $(t, t+\Delta t]$, knowing that it exists at the time t.

The expected number of occurrences of the event i in the time interval $(t, t']$ is:

$$n_i(t, t') = \int_t^{t'} \widehat{\lambda}_i(u)du.$$

Remark 5.1. The $\widehat{\lambda}(t)$ and $\widehat{\mu}(t)$ as defined above, usually called pseudo-failure and pseudo-repair rates, are different from those defined and used until now; that is, the $\lambda(t)\Delta t + o(\Delta t)$ was defined as the probability of the basic event in the time interval $(t, t+\Delta t]$, knowing that it does not exist from the instant to 0 to t. Nevertheless, in the case of an exponential law, both $\widehat{\lambda}(t)$ and $\lambda(t)$ are the same.

We proceed from the minimal cut sets as follows.

Let $K_i = \{1, ..., n_i\}$ be the ith minimal cut set, and $Q_i(t)$ its probability of occurrence at the time t. Then, $Q_i(t) = q_1(t)...q_{n_i}(t)$ where $q_i(t)$ is the probability of the basic event i at the time t.

The (pseudo-) rate of occurrence of the minimal cut set i, $\widehat{\Lambda}_i(t)$, at the time t, is:

$$\widehat{\Lambda}_i(t)\Delta t = \sum_{j=1}^{n_i} \widehat{\lambda}_i(t)\Delta t \prod_{\substack{i=1 \\ i \neq j}}^{n_i} q_i(t),$$

or

$$\widehat{\Lambda}_i(t) = \sum_{j=1}^{n_i} \widehat{\lambda}_i(t) \prod_{\substack{i=1 \\ i \neq j}}^{n_i} q_i(t).$$

The expected number of system breakdowns in the time interval $(t, t']$, due to the minimal cut set i, is:

$$N_i(t, t') = \int_t^{t'} \widehat{\Lambda}_i(u)du.$$

We will define for the system the non-availability or the probability of existence of the top event, $P_S(t)$, at the time t:

$$P_S(t) \cong \sum_{i=1}^{k} Q_i(t),$$

with k, the total number of the system's minimal cut sets.

The (pseudo-) failure rate, $\widehat{\Lambda}(t)$, of the system is:

$$\widehat{\Lambda}(t) = \sum_{i=1}^{k} \widehat{\Lambda}_i(t).$$

And the average number of the system breakdowns in the time interval $(t, t']$ is:

$$N(t, t') = \int_t^{t'} \widehat{\Lambda}(u) du.$$

5.4 Method of factorization

Another type of the probabilistic assessment of a FT, which does not necessitate the minimal sets, consists of factorizing with respect to the repeated events.

The method consists of successively factorizing the FT with respect to its repeated events. At each factorization, we have two FTs at most, containing one repeated event less. Thus, at the end of the factorization, we have a set of FTs, which do not contain a repeated event, and consequently a procedure of direct calculation can be done on them.

▷ **Example 5.2.** Factorization of the FT CH (see section 4.2): the FT CH contains two repeated events: event 3 and event 4. A successive factorization with respect to these two events gives:

$$\varphi(\mathbf{x}) = x_3 \varphi(1_3, \mathbf{x}) \dot{+} (1 - x_3) \varphi(0_3, \mathbf{x}).$$

We have: $\varphi(1_3, \mathbf{x}) = 1$. Now

$$\begin{aligned} \varphi(\mathbf{x}) &= x_3 \dot{+} (1 - x_3) \varphi(0_3, \mathbf{x}). \\ &= x_3 \dot{+} \overline{x}_3 [x_4 \varphi(0_3, 1_4, \mathbf{x}) \dot{+} \overline{x}_4 \varphi(0_3, 0_4, \mathbf{x})], \end{aligned}$$

$\varphi(0_3, 1_4, \mathbf{x}) = 1$, now

$$\varphi(\mathbf{x}) = x_3 \dot{+} \overline{x}_3 [x_4 \dot{+} \overline{x}_4 \varphi(0_3, 0_4, \mathbf{x})],$$

$\varphi(0_3, 0_4, \mathbf{x}) = x_1 x_2$

$$\varphi(\mathbf{x}) = x_3 \dot{+} \overline{x}_3 [x_4 \dot{+} \overline{x}_4 x_1 x_2].$$

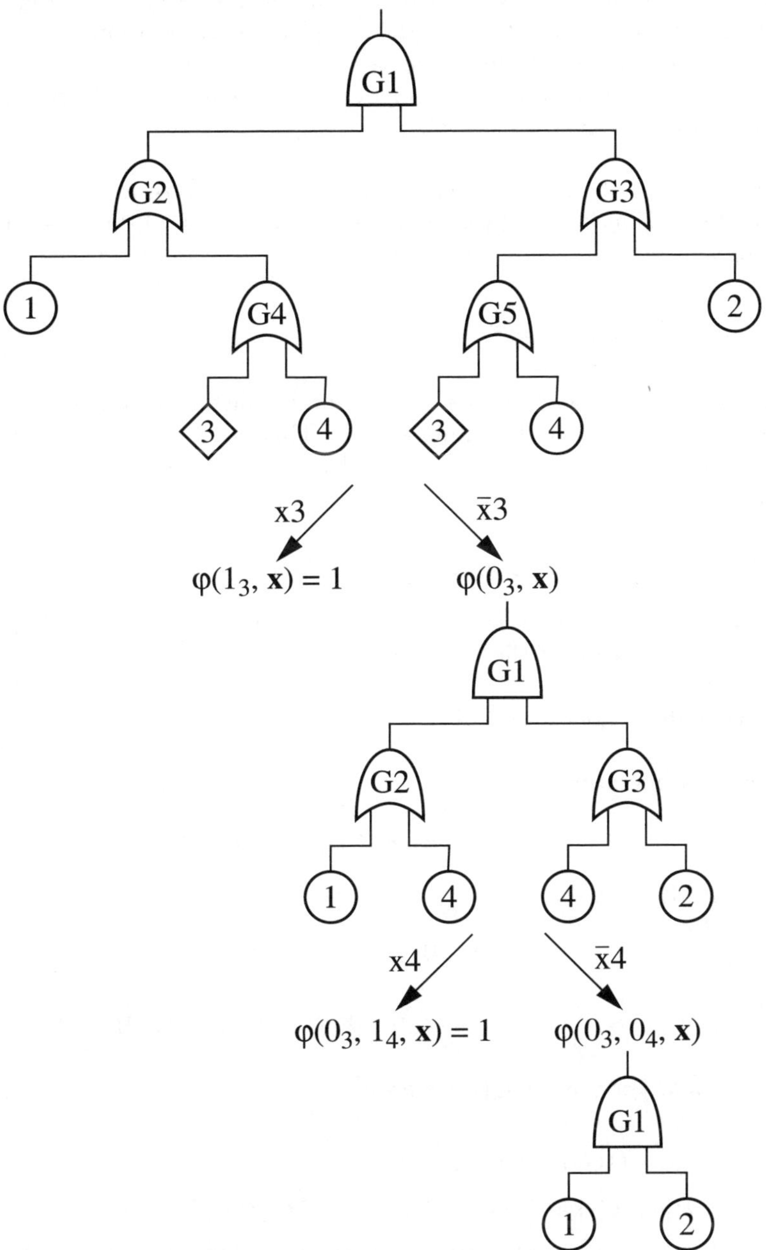

Figure 5.1 *Factorization of the FT CH*

The factorization directly carried out on the FT is given in Figure 5.1. According to Schneeweiss [SCH 84], the complete decomposition of a structure function by the Shannon formula yields an expression for the sum of disjoint products. This expression does not require the simplifications of orthogonality ($x\overline{x}=0$) and idempotance ($xx=x$).

By the same method, on stopping at a level of decomposition, we can obtain the boundaries. This is done by substituting $\varphi_i(\mathbf{x})\equiv 1$ for all the sub-structures remaining at the level, where one stops.

5.5 Direct recursive methods

The recursive methods enable a direct probabilistic assessment of the FT without passing through the minimal sets. This is important when a qualitative analysis of the FT is not required and when the exact value of the probability for the top event [LOCK 79] is calculated.

5.5.1 *Recursive inclusion-exclusion method*

The procedure is simple: we will begin from the top event and climb down the fault tree by means of the following relationships.

For two independent events A and B, we have:

For an OR operator,

$$P(A\cup B)=P(A)+P(B)-P(A)P(B).$$

For an AND operator,

$$P(A\cap B)=P(A)P(B),$$

$$P\{(A\cup B)\cap C\}=P(A\cap C)+P(B\cap C)-P(A\cap B\cap C).$$

▷ **Example 5.3.** For the FT CH, we have:

$$\begin{aligned}
Q&=P(G_2\cap G_3)=P\{(1\cup G_4)\cap G_3\}\\
&=P\{1\cap G_3\}+P\{G_4\cap G_3\}-P\{1\cap G_4\cap G_3\}\\
&=P\{1\cap(2\cup G_5)\}+P\{(3\cup 4)\cap G_3\}-P\{1\cap G_4\cap(2\cup G_5)\}\\
&=P\{1\cap 2\}+P\{1\cap G_5\}-P\{1\cap 2\cap G_5\}+P\{3\cap G_3\}\\
&\quad+P\{4\cap G_3\}-P\{3\cap 4\cap G_3\}-P\{1\cap 2\cap G_4\}\\
&\quad-P\{1\cap G_4\cap G_5\}+P\{1\cap G_4\cap 2\cap G_5\}
\end{aligned}$$

$$
\begin{aligned}
&= h_1 + P\{1 \cap (3 \cup 4)\} - P\{1 \cap 2 \cap (3 \cup 4)\} + P\{3 \cap (2 \cup G_5)\} \\
&\quad + P\{4 \cap (2 \cup G_5)\} - P\{3 \cap 4 \cap (2 \cup G_5)\} - P\{1 \cap 2 \cap (3 \cup 4)\} \\
&\quad - P\{1 \cap (3 \cup 4)\} - P\{1 \cap 2 \cap (3 \cup 4)\} \\
&= h_2 + P\{3 \cap G_5\} - P\{2 \cap 3 \cap G_5\} + P\{4 \cap G_5\} - P\{2 \cap 4 \cap G_5\} \\
&\quad - P\{3 \cap 4 \cap G_5\} + P\{2 \cap 3 \cap 4 \cap G_5\} \\
&= h_3 + P\{3 \cap (3 \cup 4)\} - P\{2 \cap 3 \cap (3 \cup 4)\} + P\{4 \cap (3 \cup 4)\} \\
&\quad - P\{2 \cap 4 \cap (3 \cup 4)\} - P\{3 \cap 4 \cap (3 \cup 4)\} \\
&\quad + P\{2 \cap 3 \cap 4 \cap (3 \cup 4)\},
\end{aligned}
$$

and on developing, we retrieve the result of section 5.1.3, where h_i, $i = 1, 2, 3$ are the terms that can be calculated; that is, the terms do not contain an operator. For example: $h_1 = P\{1 \cap 2\} = q_1 q_2$.

The advantages of a recursive algorithm, mentioned by Perry and Page [PER 86], are: exact calculation, conceptual simplicity, minimum memory capacity, and small programs.

5.5.2 *Method of recursive disjoint products*

As seen in the previous method, we climb down the FT, starting from the top event by making use of the following two relationships, for the indicator variables of the FTs events. For the basic events, we will be using the indicator variable x, and for the intermediate events, the indicator variable y. The FT is transformed as before so as to have two inputs for each operator.

Let there be an operator with two inputs, whose indicator variables are z and w.

For an OR operator:

$$z \dot{+} w = z + \overline{z} w.$$

For an AND operator:

$$zw.$$

▷ **Example 5.4.** Let us consider the FT CH.

$$\begin{aligned}\varphi(\mathbf{x}) &= y_1 = y_2y_3\\ &= y_2(x_2+\overline{x}_2y_5)\\ &= y_2x_2+y_2\overline{x}_2y_5\\ &= x_2(x_1+\overline{x}_1y_4)+\overline{x}_2y_2(x_3+\overline{x}_3x_4)\\ &= x_1x_2+\overline{x}_1x_2(x_3+\overline{x}_3x_4)+\overline{x}_2x_3y_2+\overline{x}_2\overline{x}_3x_4y_2\\ &= x_1x_2+\overline{x}_1x_2x_3+\overline{x}_1x_2\overline{x}_3x_4+\overline{x}_2x_3(x_1+\overline{x}_1y_4)\\ &\quad +\overline{x}_2\overline{x}_3x_4(x_1+\overline{x}_1y_4).\end{aligned}$$

On simplifying, we get:

$$\begin{aligned}\varphi(\mathbf{x}) = {} & x_1x_2+\overline{x}_1x_2x_3+\overline{x}_1x_2\overline{x}_3x_4+x_1\overline{x}_2x_3+\overline{x}_1\overline{x}_2x_3+x_1\overline{x}_2\overline{x}_3x_4\\ & +\overline{x}_1\overline{x}_2\overline{x}_3x_4,\end{aligned}$$

Leading to:

$$\begin{aligned}Q(\mathbf{q}) = {} & q_1q_2+p_1q_2q_3+p_1q_2p_3q_4+q_1p_2q_3+q_1p_2q_3+q_1p_2p_3q_4\\ & +p_1p_2p_3q_4.\end{aligned}$$

Remark 5.2. From the above expression for $\varphi(\mathbf{x})$, we can obtain the cut sets of the FT by eliminating the supplementary variables and, through reduction, we can obtain the minimal cut sets.

5.6 Other methods for calculating the fault trees

Apart from the methods, which we have encountered up to now, and the methods truncating the minimal sets concerning the large FTs that we are going to study in the succeeding section, there are still more methods for calculating the FTs, which we will not cover here.

One method that has been in use for a long time is the so-called method of lambda-tau (λ, τ). It leads to simple calculations and, when the FT does not contain a repeated event, we obtain the exact stationary value of (λ, τ) dealing with the top event [LIE 75] [RAC 76].

Modularization is an effective method in combination with other methods (for example direct calculation, minimal sets, etc.), as it transforms the calculation of an FT to the calculation of small independent FTs (cf. section 7.2).

When we have $\lambda_i(t) = K_i t_i^m$ (Weibull law), $i = 1, ..., N$, we obtain, by means of the minimal cut sets of the FT, the $S(t)$ of the top event through a method proposed in [DUB 80].

The calculation of the boundaries, presented in Chapter 2, is also very much in use in combination with the modularization.

5.7 Large fault trees

One of the important limitations concerning the applications of the FTs is the large number of the minimal sets that they generate. We quickly reach some thousands and even millions of minimal cut sets.

The treatment of such an FT, even for calculating the lower boundary, is quite forbidding.

Because of this fact, the methods of truncating the set of minimal cut sets, developed in the 1980s, have certain importance. The basic idea of these methods consists of eliminating the minimal cut sets, whose contribution to the probability of the top event is below a given threshold, or the minimal cut sets are of a length exceeding a given value.

An assessment of the error that occurred is given with each of these methods. We will be presenting the principal methods of truncation.

5.7.1 *Method of Modarres and Dezfuli [MOD 84]*

Let $q_1, ..., q_N$ be the probabilities of the basic evens of the FT with $q = \max\{q_1, ..., q_N\}$ and n_k be the number of minimal cut sets of length k, $1 \leq k \leq N$. We have:

$$n_k \leq \Delta = C_k^N = \frac{N!}{(N-k)!k!}.$$

For very large values of N, we have the approximation:

$$\Delta \cong \frac{N^k}{k!}.$$

The probability of the occurrence of the minimal cut sets of length k, marked as P_k, is:

$$P_k = P\{K_{(1)} \cup ... \cup K_{(n_k)}\} \leq 1 - [1 - P(K_{(1)})]...[1 - P(K_{(n_k)})],$$

whence,

$$P_k \leq P_k' = 1 - (1 - q^k)^{\Delta}.$$

We have: $P_k' \to 1 - \exp(-\nu^k/k!)$, when $N \to \infty$, where $\nu = qN$.

If $\nu < 1$, the contribution of the minimal cut sets to the top event rapidly decreases, and the contribution of the cut sets of length k is less important than that of the cut sets of length $k-1$. If $\nu > 1$, then one has to break up the FT into sub-FTs having the values $\nu < 1$.

5.7.2 *Method of Hughes [HUG 87]*

This method is a variant of the previous method, and the difference between them lies in the calculation of the probability of the cut sets of length k, P_k'.

$$P_k' = \sum_{\substack{r_i=0,1 \\ r_1+\ldots+r_N=k}} (1-q_1)^{1-r_1} q_1^{r_1} \ldots (1-q_N)^{1-r_N} q_N^{r_N}.$$

Let:

$$h_k = A \frac{\alpha^k}{k!},$$

with

$$A = \prod_{i=1}^{N} (1 - q_i) \quad \text{and} \quad \alpha = \sum_{i=1}^{N} \frac{q_i}{1 - q_i}.$$

We have:

$$P_k' \leq h_k.$$

A bound of the error committed by omitting the minimal cut sets of the order higher than ω is:

$$\beta(\omega + 1) = \beta(0) - \sum_{k=1}^{\omega} h_k$$

where $\beta(0) = Ae^{\alpha}$.

In the case where the distribution of the number of cut sets in terms of orders of length is not regular, or where the values of probability of the basic events are very much dispersed, the value of P_k' will be very much above the real value (for example, 1,000% more important than the real value of the error!).

The improvements that have been brought about [KER 91] are of two kinds. They are concerned on the one hand with the use of the value of the maximum number of cut sets and the maximum length of the cut sets instead of the number N (these numbers being very rapidly obtained on the FT through the algorithms of linear complexity), and on the other hand with the use of Shannon's formula (cf. relation (2.5)).

5.7.3 *Schneeweiss method [SCH 87]*

Let φ be the structure function of the FT written under the form of a sum of products corresponding to the minimal cut sets. We can write it as:

$$\varphi = \varphi_1 \dot{+} \varphi_2.$$

Proceeding from this equality, and by applying the expectation operator, we have:

$$E\varphi_1 \leq E\varphi \leq E\varphi_1 + E\varphi_2.$$

Now, the maximum error committed, namely $\varepsilon_{\max}$, by omitting the term φ_2, is less than or equal to $E\varphi_2$. In addition, we have, if $\varphi_2 = y_1 \dot{+} \cdots \dot{+} y_s$:

$$E\varphi_2 \leq EY_1 + ... + EY_s,$$

Consequently:

$$\varepsilon_{\max} \leq EY_1 + ... + EY_s.$$

The algorithm of Schneeweiss for determining the φ_2 and the algorithms that follow from this analysis are as follows:

Algorithm 5.2.

1. Fix a maximum error: $\varepsilon_{\max}$;
2. Calculate the $EY_1, ..., EY_k$ and arrange them in increasing order: $EY_{(1)} \leq ... \leq EY_{(k)}$;
3. Calculate successively: $EY_{(1)}, EY_{(1)} + EY_{(2)}, ...$, until we get: $\varepsilon_{\max} < EY_{(1)} + ... + EY_{(J)}$, then put: $\varepsilon = EY_{(1)} + ... + EY_{(J-1)}$;
4. $\varphi_2 = y_{(1)} \dot{+} ... \dot{+} y_{(J-1)}$;

End.

5.7.4 *Brown method [BRO 90]*

This method consists of climbing up the FT by obtaining the minimal cut sets whose probability of individual occurrence is higher than a fixed value ε. Let the intermediate event E and $\mathcal{K}_E$ be the set of the minimal cut sets of E. Let us consider the following partition:

$$\mathcal{K}_E = \mathcal{K}_{E,1} \cup \mathcal{K}_{E,2}.$$

For any minimal cut set X of $\mathcal{K}_{E,1}$, we have: $P\{X\} > \varepsilon$, and for any minimal cut set Y of $\mathcal{K}_{E,2}$ we have: $P\{Y\} \leq \varepsilon$.

Thus, we retain the minimal cut sets $\mathcal{K}_{E,1}$ and eliminate the $\mathcal{K}_{E,2}$ on having a probability of error ε_E corresponding to the eliminated cut sets:

$$P\{X \cup Y\} \leq \varepsilon_E = P\{X\} + P\{Y\} - P\{X\}P\{Y\}. \quad (5.1)$$

Two cases appear according to the operator corresponding to the event E.

OR operator: $E = X \cup Y$

$$\varphi_E = (\varphi_{X,1} \dot{+} \varphi_{Y,1}) \dot{+} (\varphi_{X,2} \dot{+} \varphi_{Y,2}),$$

whence: $\mathcal{K}_{E,1} = \mathcal{K}_{X,1} \cup \mathcal{K}_{Y,1}$

$$E(\varphi_{X,2} \dot{+} \varphi_{Y,2}) \le \varepsilon_E = E(\varphi_{X,2}) + E(\varphi_{Y,2}) - E(\varphi_{X,2})E(\varphi_{Y,2}).$$

AND operator: $E = X \cap Y$

$$\begin{aligned}\varphi_E &= (\varphi_{X,1} \dot{+} \varphi_{X,2})(\varphi_{Y,1} \dot{+} \varphi_{Y,2}) \\ &= (\varphi_{X,1}\varphi_{Y,1}) \dot{+} \{(\varphi_{X,1}\varphi_{Y,2}) \dot{+} \varphi_{X,2}(\varphi_{Y,2} \dot{+} \varphi_{Y,1})\}\end{aligned}$$

Thus:

$$\mathcal{K}_{E,1} \subset \mathcal{K}_{X,1} \cap \mathcal{K}_{Y,1}. \tag{5.2}$$

The calculation is done by the Brown algorithm as follows.

Algorithm 5.3.

1. Obtain the minimal cut sets from (5.2).
2. Calculate the upper boundary of $P\{(\mathcal{K}_{X,1} \cap \mathcal{K}_{Y,1}) \setminus \mathcal{K}_{E,1}\}$.
3. Upper boundary of $P\{\mathcal{K}_{X,1} \cap \mathcal{K}_{Y,2}\}$ is equal to the min $\{b_s(P\{\mathcal{K}_{X,1}\}), P\{\mathcal{K}_{Y,2}\}\}$.
4. Upper boundary of $P\{\mathcal{K}_{X,2} \cap (\mathcal{K}_{Y,2} \cup \mathcal{K}_{Y,1})\} = \min\{b_s(P(\mathcal{K}_{X,1})), b_s(P\{\mathcal{K}_{Y,1} \cup \mathcal{K}_{Y,2}\})\}$, where $b_s(\cdot)$ is the upper boundary of $(\cdot)$ obtained through relationship (5.1).

End.

Chapter 6

Influence Assessment

6.1 Uncertainty

6.1.1 *Introduction*

In the previous chapter, we studied the probability of the top event considering that the parameters of distribution concerning the basic events were fixed.

In reality, this is not always the case. There are very good reasons for saying that this approach is not always satisfactory. In fact, the data on the reliability of components are generally obtained through tests or even through field data. In both cases, a statistical analysis is carried out in order to determine the parameters of distributions, wherein we consider these parameters that constitute themselves as random variables. Hence, for every parameter, we define a distribution, or the first moments, or just a simple factor of error, etc.

Concerning a component, whose lifetime follows a distribution depending on a parameter, we obtain, within the framework of a standard statistical study, an estimator (preferably an estimator with good properties!), whose distribution enables us to obtain confidence intervals of the parameter at desired levels of confidence, and then the confidence intervals for the probability of this component's failure. Obviously an analysis with the same aim as that of the preceding analysis can be carried out within a Bayesian analysis framework.

Within the framework of the reliability of the systems, and in particular, in matters concerning the FTs, the previous analysis dealing with the components of the system is considered a data, and we seek to start from there to obtain an interval of confidence for the probability of the top event.

The fact of considering the parameters of a distribution as random variables is due not only to the uncertainty about the data, but also on account of the mode of using the system. To illustrate this, we give an example [BIE 83].

▷ **Example 6.1.** Let us consider a system with a single component meant for working over the time interval $[0, \theta]$. The component will be picked up at random from a stock of components coming from two different stocks, noted as A and B. The stock is constituted by n components in A and by m components in B. The probabilities of the failure of components over $[0, \theta]$ are p_1 for a component A, and p_2 for a component B.

The probability that the system breaks down over $[0, \theta]$ is a random variable, whose mathematical expectation is given by:

$$E(P_S) = p_1 \frac{n}{n+m} + p_2 \frac{m}{n+m}.$$

6.1.2 *Methods for evaluating the uncertainty*

There are two large families of methods that are used for evaluating the uncertainty of the top event: analytical methods and numerical methods.

The analytical methods consist of evaluating a few first moments of distribution of the top event from the first moments of the basic events. With the help of moments evaluated for the top event, we either make use of the inequalities so as to deduce from them the confidence intervals, or we adopt a distribution.

In the case of numerical methods, it is assumed that the distributions of the basic events are known, from which we evaluate the distribution of the top event, either through the Monte-Carlo method or through the method of discrete distributions.

In this section, we discuss the analytical methods developed by Apostolakis *et al.* [APO 77].

Inequality methods

Inequality methods are very often useful and are valid for any distribution, and therefore they can be used without knowing the distribution except some of their first moments, usually the average and the variance.

– Markov inequality: if X is a real positive random variable, then for all real $\delta > 0$ we have:

$$P[X \geq \delta] \leq \frac{E[X]}{\delta}.$$

– Bienaymé-Chebychev inequality: if X is a random variable of the average μ and of the variance σ^2, then we have:

$$P[|X - \mu| \geq \delta] \leq \frac{\sigma^2}{\delta^2}.$$

More generally, if X has moments of the order $n \geq 1$, we have:

$$P[|X - \mu| \geq \delta] \leq \frac{E[|X - \mu|^n]}{\delta^n}.$$

Method of empirical distributions [APO 77]

This consists of using a distribution for the random variable P_S. The majority of the distributions that were used have the disadvantage of the domain of definition being much larger than the interval $[0, 1]$. The empirical distribution of Johnson S_B can, on the other hand, be defined over $[0, 1]$, which is coherent with the nature of P_S. The random variable P_S follows a Johnson S_B distribution if

$$f.r.(P_S) = \Phi\Big(\frac{\ln \frac{P_S}{1-P_S} - \mu}{\sigma}\Big),$$

where $\Phi(\cdot)$ is the standard Gaussian c.d.f.

Once the μ_S and σ_S are evaluated, we have to define the μ and σ of a Gaussian r.v., by resolving the system of two equations as follows:

$$\int_0^1 x d_x[\Phi\Big(\frac{\ln \frac{x}{1-x} - \mu}{\sigma}\Big)] = \mu_S,$$

$$\int_0^1 (x - \mu_S)^2 d_x[\Phi\Big(\frac{\ln \frac{x}{1-x} - \mu}{\sigma}\Big)] = \sigma_S^2.$$

6.1.3 *Evaluation of the moments*

Certain methods have been proposed in other works for evaluating the moments (mainly the μ_S and the σ_S^2).

Direct method

For a direct evaluation on the FT, we can make use of the following relationships.

For an OR operator, we have:

$$E(P_S) = E(P_A) + E(P_B) - E(P_A)E(P_B),$$

$$Var(P_S) = Var(P_A)[1 - E(P_B)]^2 + Var(P_B)[1 - E(P_A)]^2 + Var(P_A)Var(P_B).$$

For an AND operator, we have:

$$E(P_S) = E(P_A)E(P_B),$$

$$Var(P_S) = Var(P_A)[E(P_B)]^2 + Var(P_B)[E(P_A)]^2 + Var(P_A)Var(P_B).$$

▷ **Example 6.2.** Let the FT be as given in Figure 6.1.

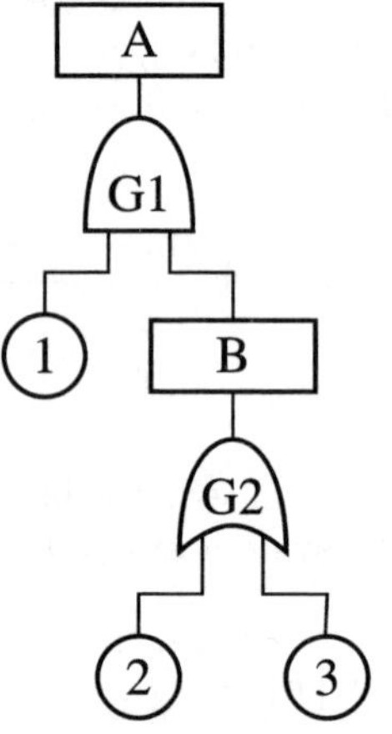

Figure 6.1 *FT of example 6.2*

with the following data:

Component	Mean value	Variance
1	10^{-2}	10^{-3}
2	10^{-3}	10^{-4}
3	5×10^{-3}	2×10^{-4}

We have:

Mean value:

$$\mu_B = \mu_2 + \mu_3 - \mu_2\mu_3, \quad \mu_B = 5.995 \cdot 10^{-3}$$

$$\mu_A = \mu_1\mu_B, \quad \mu_S = \mu_A = 5.995 \cdot 10^{-4};$$

Variance:

$$\sigma_B^2 = \sigma_2^2[1-\mu_3]^2 + \sigma_3^2[1-\mu_2]^2 + \sigma_2^2\sigma_3^2, \quad \sigma_B^2 = 2.986 \cdot 10^{-4}$$

$$\sigma_A^2 = \sigma_1^2\sigma_B^2 + \sigma_1^2\mu_B^2 + \sigma_B^2\mu_1^2, \quad \sigma_S^2 = \sigma_A^2 = 3.64 \cdot 10^{-7}.$$

Rushdi's method

The method developed by Rushdi [RUS 85] is based on the complete development into the Taylor series of the function $Q(\mathbf{q})$ for the top event. As this function is a multiaffine function, it has a finite Taylor series development. This development, around the mean, is as follows:

$$Q(\mathbf{q}) - Q(\mathbf{m}_1) = \sum_{i=1}^{n} C_i(q_i - m_{i1}) + \sum_{i=1}^{n-1}\sum_{j=i}^{n} C_{ij}(q_i - m_{i1})(q_j - m_{j1})$$
$$+ ... + C_{12...n}(q_1 - m_{11})(q_2 - m_{21})...(q_n - m_{n1}) \qquad (6.1)$$

where $m_{in} = E[|X_i - \mu_i|^n]$ is the random variable X_i's n^{th} centered moment. The coefficients $C_i, C_{ij}, \ldots, C_{12\ldots n}$ are calculated from the following relationships:

$$Q(\mathbf{q}) = q_i Q(1_i, \mathbf{q}) + (1 - q_i)Q(0_i, \mathbf{q}).$$

We have:

$$C_i = (\frac{\partial Q}{\partial q_i})_{\mathbf{q}=\mathbf{m}_1} = Q(1_i, \mathbf{m}) - Q(0_i, \mathbf{m}),$$

$$C_{ij} = (\frac{\partial^2 Q}{\partial q_i \partial q_j})_{\mathbf{q}=\mathbf{m}_1} = Q(1_i, 1_j, \mathbf{m}_1) - Q(0_i, 1_j, \mathbf{m}_1)$$
$$-Q(1_i, 0_j, \mathbf{m}_1) + Q(0_i, 0_j, \mathbf{m}_1),$$

$$C_{ijk} = (\frac{\partial^3 Q}{\partial q_i \partial q_j \partial q_k})_{\mathbf{q}=\mathbf{m}_1}$$
$$= Q(1_i, 1_j, 1_k, \mathbf{m}_1) - Q(0_i, 1_j, 1_k, \mathbf{m}_1)$$
$$-Q(1_i, 0_j, 1_k, \mathbf{m}_1) + Q(0_i, 0_j, 1_k, \mathbf{m}_1)$$
$$-Q(1_i, 1_j, 0_k, \mathbf{m}_1) + Q(0_i, 1_j, 0_k, \mathbf{m}_1)$$
$$+Q(1_i, 0_j, 0_k, \mathbf{m}_1) - Q(0_i, 0_j, 0_k, \mathbf{m}_1).$$

From relationship (6.1), developed to the power n, and by applying the expectation operator, we can obtain the moments of order n.

▷ **Example 6.3.** The structure function of the FT is:

$$\varphi(\mathbf{x}) = x_1 x_2 \dot{+} x_1 x_3,$$

hence

$$Q(\mathbf{q}) = q_1 q_2 + q_1 q_3 - q_1 q_2 q_3.$$

The development of Taylor for $Q(\mathbf{q})$ around the mean is:

$$\begin{aligned} Q(\mathbf{q}) - Q(\mathbf{m}_1) = {} & C_1(q_1 - m_{11}) + C_2(q_2 - m_{21}) + C_3(q_3 - m_{31}) \\ & + C_{12}(q_1 - m_{11})(q_2 - m_{21}) + C_{13}(q_1 - m_{11})(q_3 - m_{31}) \\ & + C_{23}(q_2 - m_{21})(q_3 - m_{31}) \\ & + C_{123}(q_1 - m_{11})(q_2 - m_{21})(q_3 - m_{31}). \end{aligned}$$

The coefficients are:

$$\begin{aligned} C_1 &= q_2 + q_3 - q_2 q_3, \\ C_2 &= q_1 - q_1 q_3, \\ C_3 &= q_1 - q_1 q_2, \\ C_{12} &= 1 - q_3, \\ C_{12} &= 1 - q_2, \\ C_{12} &= -q_1, \\ C_{123} &= 1. \end{aligned}$$

The variance for the probability of the top event is:

$$\begin{aligned} \sigma_S^2 = {} & [m_{21} + m_{31} - m_{21} m_{31}]^2 m_{12} + m_{11}^2 [1 - m_{31}]^2 m_{22} \\ & + m_{11}^2 [1 - m_{21}]^2 m_{32} + [1 - m_{31}]^2 m_{12} m_{22} \\ & + [1 - m_{21}]^2 m_{12} m_{32} + m_{11}^2 m_{22} m_{32} + m_{12} m_{22} m_{32}. \end{aligned}$$

6.2 Importance

6.2.1 *Introduction*

Until now, we have been focusing on the evaluation of the probability of the top event in a coherent FT when the probabilities of the basic events are known.

Now, there arises another important question regarding the particular role played by a basic event from the viewpoint of its contribution to the probability of the top event. During the stage of designing of a system, before defining the technical specifications for the components, we do not have any probabilistic data. In view of the ruggedness of the structure (see Figure 6.2), we are tempted to conclude that component 1 is the most important (the most sensitive from the viewpoint of good functioning of the system). In other words, the failure in this component leads, on its own, to the failure of the system, which is not the case for components 2 and 3. The failure in component 2 cannot, on its own, lead to the failure of the system. This type of approach, in the absence of probabilistic data, concerns the importance of the component from the structural point of view.

In a second stage, once the technical specifications have been defined and we have the probabilistic data, the context of our approach is changed; that is, we will be asking questions about the probabilistic importance of the components. In other words, how much will the reliability of a component contribute to the reliability of the system? Or, what is the probability that the component i would have brought about the failure of the system? etc.

This type of study guides us in optimizing an investment with a view to improving the reliability of the system.

The values used for measuring the importance of the components are called the “factors of importance”.

It must be noted that apart from the components, we can ask ourselves the same questions for a particular module or even for a minimal set. Analogous measures exist for the latter.

Here is an example that clarifies this question.

▷ **Example 6.4.** Let us consider the system given in Figure 6.2.

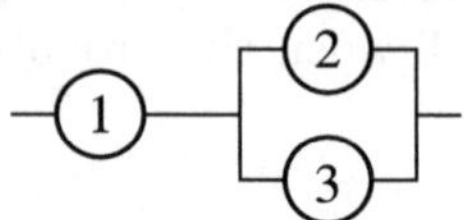

Figure 6.2 *Binary system of order 3*

Let us note by $p_i(t)$ the reliability of the component i $(i = 1, 2, 3)$ and $\mathbf{p} = [p_1, p_2, p_3]$, and $R(t) = r(\mathbf{p}(t))$ the reliability of the system; $r(\mathbf{p})$ is the reliability of structure. We can write:

$$\frac{dR}{dt}(t) = \frac{\partial r}{\partial p_1}\frac{dp_1}{dt} + \frac{\partial r}{\partial p_2}\frac{dp_2}{dt} + \frac{\partial r}{\partial p_3}\frac{dp_3}{dt},$$

or, for small variations, we write:

$$\Delta R \cong \frac{\partial r}{\partial p_1}\Delta p_1 + \frac{\partial r}{\partial p_2}\Delta p_2 + \frac{\partial r}{\partial p_3}\Delta p_3.$$

The reliability of the system is given by:

$$R(t) = p_1(t)p_2(t) + p_1(t)p_3(t) - p_1(t)p_2(t)p_3(t), \tag{6.2}$$

and consequently,

$$a_1 = \frac{\partial r}{\partial p_1} = p_2 + p_3 - p_2 p_3,$$

$$a_2 = \frac{\partial r}{\partial p_2} = p_1 - p_1 p_3,$$

$$a_3 = \frac{\partial r}{\partial p_3} = p_1 - p_1 p_2.$$

Numerical application: let $p_1 = p_2 = p_3 = 0.9$, then we have: $a_1 = 0.99$ and $a_2 = a_3 = 0.099$; and $R = 0.891$.

If we increase the reliability of the component 1, such that $p_1 = 0.95$ and the reliability of the other components remaining the same, we have:

$$\Delta R \cong 0.99 \cdot (0.95 - 0.90) = 0.0495.$$

The reliability given by the relationship (6.2) will be:

$$R = 0.9405 \quad \text{and} \quad \Delta R \cong 0.9405 - 0.891 = 0.0495.$$

If at present, we increase the reliability of the component 2 such that $p_1 = 0.95$ and the reliability of the other components remaining the same, we have:

$$\Delta R \cong 0.099 \cdot (0.95 - 0.90) = 0.00495.$$

Then, for the same increase in the reliabilities of components 1 and 2, the reliability of the system will not increase in the same manner; far from it!

6.2.2 *Structural importance factors*

Before giving the factors of structural importance, we give the following definition:

Critical vector for the component i: the vector $\mathbf{x}$ is said to be critical for the component i, if $\varphi(1_i, \mathbf{x}) = 1$ and $\varphi(0_i, \mathbf{x}) = 0$.

▷ **Example 6.5.** Let us consider the system as given in Figure 6.2. The vectors $(1,1,0)$, $(1,0,1)$ and $(1,1,1)$ are critical vectors for component 1. The vector $(1,1,0)$ is critical for component 2.

The number of critical vectors for component i, noted as $n_\varphi(i)$, is given by the relationship:

$$n_\varphi(i) = \sum_{\{\mathbf{x}|x_i=1\}} [\varphi(1_i, \mathbf{x}) - \varphi(0_i, \mathbf{x})].$$

We give the two factors of structural importance that were used.

Birnbaum's factor of importance: $I_B^\varphi(i)$

This is expressed by the ratio of the number of critical vectors for the component i to the total number of vectors [BAR 75].

$$I_B^\varphi(i) = \frac{\partial r}{\partial p_i}(\mathbf{1/2}) = \frac{1}{2^{n-1}} n_\varphi(i),$$

where $\mathbf{1/2} = (1/2, ..., 1/2)$.

Barlow-Proschan's factor of importance: $I_{BP}^{\varphi}(i)$

This is expressed by the "average probability" of critical vectors for the event i. Its expression is given by the relationship:

$$I_{BP}^{\varphi}(i) = \int_0^1 [Q(1_i, \mathbf{q}) - Q(0_i, \mathbf{q})]dq,$$

with $\mathbf{q} = (q, ..., q)$.

It can be calculated from:

$$I_{BP}^{\varphi}(i) = \frac{1}{n}\sum_{k=1}^{n}\frac{n_{\varphi}(i)}{C_{k-1}^{n-1}}.$$

6.2.3 *Probabilistic importance factors*

Birnbaum's importance factor: $I_B^{(i)}$

This is expressed by the probability that the vector $X(t)$ would be critical for the event i at the time t.

$$I_B^{(i)}(t) = \frac{\partial Q}{\partial q_i}(\mathbf{q}(t)) = Q(1_i, \mathbf{q}(t)) - Q(0_i, \mathbf{q}(t)).$$

Criticality importance factor: $I_C^{(i)}$

This is expressed by the probability that the vector $X(t)$ would be critical for the event i at the time t, and that the top event could exist at the time t.

$$I_C^{(i)}(t) = \frac{Q(1_i, \mathbf{q}(t)) - Q(0_i, \mathbf{q}(t))}{Q(\mathbf{q}(t))}q_i(t).$$

Vesely-Fussell's importance factor: $I_{VF}^{(i)}$

This is expressed by the probability that the event i would have contributed to the occurrence of the top event, knowing that this took place prior to t.

$$I_{VF}^{(i)}(t) = \frac{P\{U_i(t) = 1\}}{Q(\mathbf{q}(t))},$$

where $U_i(t) = 1 - \prod_{j=1}^{N_i}[1 - \prod_{k \in K_j} X_k(t)]$ represents the Boolean sum of the minimal cut sets containing the event i, and N_i represents the number of minimal cut sets containing component i.

Lambert's importance factor of improvement: $I_{La}^{(i)}$

This is expressed by the percentage rate of change in the probability of the top event resulting from a percentage change in the probability of the event i at the time t.

$$I_{La}^{(i)}(t) = \lambda_i(t) \frac{\partial}{\partial \lambda_i(t)} h(\lambda(t)),$$

where $h = Q \circ g$ and $g \colon \lambda \mapsto \mathbf{q}$ and $\lambda = (\lambda_1, ..., \lambda_n)$ the failure rates of components.

We can write this factor of importance by replacing, in the formula mentioned above, λ_i with $\gamma_i = \lambda_i / \lambda_{ref}$ where λ_{ref} is a reference value:

$$\gamma_i(t) \frac{\partial}{\partial \gamma_i(t)} Q(\gamma, \mathbf{q}(t)),$$

or even as:

$$\frac{\gamma_i(t)}{Q(\gamma, \mathbf{q}(t))} \frac{\partial}{\partial \gamma_i(t)} Q(\gamma, \mathbf{q}(t)).$$

These last two expressions give the same grading for the basic events, but the second expression gives values closer to unity.

Factor of importance for the sequential contribution according to Lambert: $I_{Ls}^{(i)}$

This is expressed by the probability that the vector $\mathbf{X}(t)$ would be critical for event i at time t and that another event triggers the occurrence of the top event.

$$I_{Ls}^{(i)}(t) = \frac{1}{Q(\mathbf{q}(t))} \sum_{j:j \neq i} \int_0^t [Q(1_i, 1_j, \mathbf{q}(t)) - Q(1_i, 0_j, \mathbf{q}(t))] q_j(u) df_j(u).$$

Barlow-Proschan's importance factor: $I_{BP}^{(i)}$

This is expressed by the probability that the event i would have triggered the occurrence of the top event.

$$I_{BP}^{(i)} = \int_0^\infty [Q(1_i, \mathbf{q}(t)) - Q(0_i, \mathbf{q}(t))] dq_i(t).$$

We have $0 \leq I_{BP}^{(i)} \leq 1$ and $\sum_{i=1}^n I_{BP}^{(i)} = 1$.

Natvig's importance factor: $I_N^{(i)}$

It is defined by means of the random variable Z_i which designates the reduction of the residual time of the top event's occurrence due to the occurrence of the event i.

Let:

L_i^1 be the residual lifetime of the system just before the failure in the component i

L_i^0 be the residual lifetime of the system just after the failure in the component i.

We have:

$$Z_i = L_i^1 - L_i^0.$$

Natvig's factor of importance is expressed by the following relationship:

$$I_N^{(i)} = \frac{E[Z_i]}{\sum_{j=1}^n E[Z_j]}.$$

It is clear that we have: $0 \leq I_N^{(i)} \leq 1$ and $\sum_{i=1}^n I_N^{(i)} = 1$.

The expectation $E[Z_i]$ is calculated as follows:

$$E[Z_i] = E[L_i^1] - E[L_i^0] = \int_0^\infty P\{L_i^1 > x\}dx - \int_0^\infty P\{L_i^0 > x\}dx,$$

and

$$\int_0^\infty P\{L_i^1 > x\}dx = \int_0^\infty \sum_{(\cdot_i,\mathbf{x})} \prod_{j\neq i} (F_i(t))^{1-x_j}(\overline{F}_i(t))^{x_j} f_i(t) Q(\overline{H}_t^{(1_i,\mathbf{x})}(u))dt,$$

$$\int_0^\infty P\{L_i^0 > x\}dx = \int_0^\infty \sum_{(\cdot_i,\mathbf{x})} \prod_{j\neq i} (F_i(t))^{1-x_j}(\overline{F}_i(t))^{x_j} f_i(t) Q(\overline{H}_t^{(0_i,\mathbf{x})}(u))dt,$$

where

$$\overline{H}_t^{\mathbf{x}}(u) = (\overline{H}_{(1,t)}^{x_1}(u), ..., \overline{H}_{(n,t)}^{x_n}(u))$$

and

$$H_{(i,t)}^1(u) = \frac{\overline{F}_i(t+u)}{\overline{F}_i(t)}, \quad \overline{H}_{(n,t)}^0(u) = 0.$$

6.2.4 *Importance factors over the uncertainty*

For completing the study of uncertainty, we will be studying the importance of the basic events on the uncertainty of the top event's probability. In particular, we will be presenting the factors of importance concerning the influence of the basic event probability variances on the top event probability variance.

We will be presenting two factors of importance over the uncertainty; one of them is introduced by Pan and Tai [PAN 88], which is analogous to that of probabilistic factor of importance of Birnbaum. The other is introduced by Bier [BIE 83], which is analogous to the improvement factor according to Lambert.

The probabilities of the basic events q_i are random variables, and consequently the probability function of the top event $Q(\mathbf{q})$ is a random variable too.

Pan-Tai's factor of importance: $\Pi_{PT}^{(i)}$

Analogous to the factor of importance of Birnbaum, this is defined by:

$$\Pi_{PT}^{(i)} = \frac{\partial Var(Q)}{\partial Var(q_i)}.$$

It can be expressed by:

$$\Pi_{PT}^{(i)} = E\left(\frac{\partial Q}{\partial q_i}\right)^2.$$

Bier's factor of importance: $\Pi_{BR}^{(i)}$

Analogous to the factor of importance by Lambert, this is defined by:

$$\Pi_{BR}^{(i)} = \frac{Var(q_i)}{Var(Q)} \frac{\partial Var(Q)}{\partial Var(q_i)}.$$

We have:

$$0 \leq \Pi_{BR}^{(i)} \leq 1, \quad i = 1, ..., n.$$

▷ **Example 6.6.** Let us consider the FT given in Figure 6.1. The probability variance of the top event is given by the relationship (see section 6.1.3).

$$\begin{aligned} Var(Q) = {} & [m_{21} + m_{31} - m_{21}m_{31}]^2 m_{12} + m_{11}^2 [1 - m_{31}]^2 m_{22} \\ & + m_{11}^2 [1 - m_{21}]^2 m_{32} + [1 - m_{31}]^2 m_{12} m_{22} \\ & + [1 - m_{21}]^2 m_{12} m_{32} + m_{11}^2 m_{22} m_{32} + m_{12} m_{22} m_{32}. \end{aligned}$$

Hence,

$$\mathit{\Pi}_{PT}^{(i)} = [m_{21} + m_{31} - m_{21}m_{31}]^2 + [1 - m_{31}]^2 m_{22} + [1 - m_{21}]^2 m_{32} + m_{22}m_{32},$$

and

$$\mathit{\Pi}_{BR}^{(i)} = [[m_{21} + m_{31} - m_{21}m_{31}]^2 + [1 - m_{31}]^2 m_{22} + [1 - m_{21}]^2 m_{32} + m_{22}m_{32}]m_{12}/var(Q).$$

Chapter 7

Modules – Phases – Common Modes

7.1 Introduction

In this chapter, we will study the modular decomposition of the FTs, the FTs with phases and the common failure modes.

These questions are important for several reasons. The modular decomposition of an FT enables an easier calculation of the probability of the top event or the calculation for better bounds. In addition, it can lead to a better understanding of the tree structure and, consequently, carry out a better qualitative analysis.

The FTs with phases enable us to carry out an overall study of the systems in phases, that is, the systems having many phases functioning during the currency of their mission. For example, an aircraft has three working phases during the mission of a flight from one place to another: the takeoff phase, the flight phase and the landing phase. The FT with phases concerning the failure of the mission represents the three phases of flight, and its treatment is global.

Finally, the multiple failures, which are a result of common causes that wipe out the advantage of the redundancies and other functional dependences under the name "common mode of failures" (CM), should be seriously taken into account during dependability studies.

7.2 Modular decomposition of an FT

7.2.1 *Module and better modular representation*

The concept of module in the context of the binary systems is defined by Birnbaum and Esary in 1965 [BIR 65], Chapter 2.2, [BAR 75].

A sub-FT is said to be independent if no elements of its domain appears anywhere else in the FT.

An independent sub-FT is a module.

▷ **Example 7.1.** Let us consider the FT in Figure 7.1.

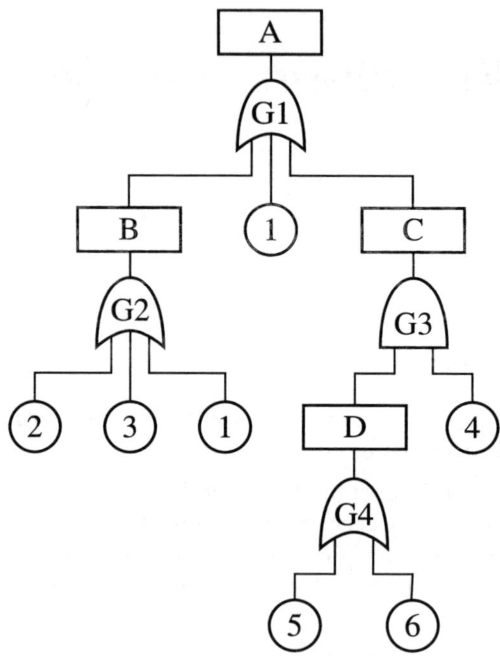

Figure 7.1 *Fault tree*

The sub-FT C is independent as no element in its domain $\{4, 5, 6\}$ appears elsewhere in the FT. On the contrary, the sub-FT B is not independent, because the event 1 appears elsewhere.

Consequently, the sub-FT C is a module of the FT A.

The best modular representation of an FT is an equivalent FT with the following properties:

(i) Reduced representation, that is to say, each operator is:

– either a primary operator,

– or an AND operator that does not have an AND operator as input,

– or an OR operator that does not have an OR operator as input.

(ii) All the sub-FTs are independent.

▷ **Example 7.2.** The best modular representation of the FT in Figure 7.1 is given in Figure 7.2.

The modularization of an FT presents numerous advantages, the most important being as follows:

– Exact evaluation of the probability of the top event through procedures that are extremely simple.

– Better bounds for the top event and an easier attainment.

– Treatment of certain statistical dependences among the basic events.

– From the best modular representation of an FT, we can obtain all its modular representations, and thus the optimum modularization can be obtained.

– When the modularization is directly carried out, that is, without passing through the minimal sets, the modularization reduces the complexity of the search for the latter.

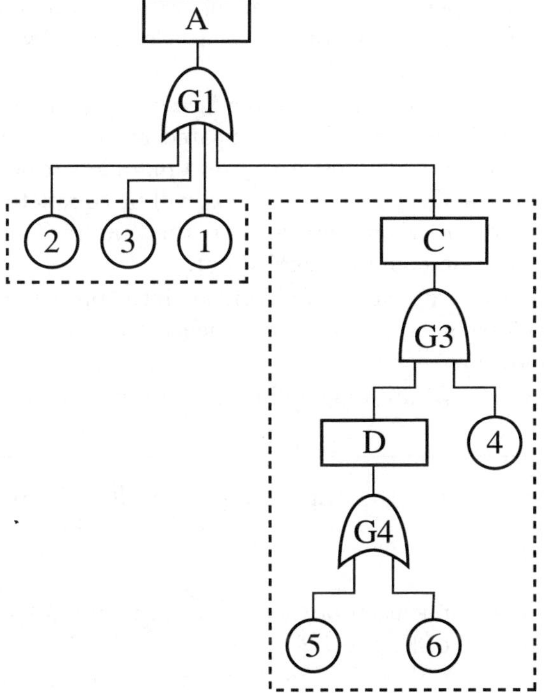

Figure 7.2 *Best modular representation*

7.2.2 *Modularization of a fault tree*

There are two approaches for the modularization of the FTs. The first approach consists of intervening before obtaining the minimal sets based on the FT, and in the other approach, the intervention comes later and is based on the structure function of the FT. With regard to the first approach, we will be presenting the algorithm by Olmos and Wolf [OLM 78]. As for the second approach, we will present a technique proposed by Wilson [WIL 86].

Algorithm by Olmos and Wolf

This algorithm is presented in six stages:

Algorithm 7.1.

1. The vertices having repeated events in common are interconnected. These interconnections define the set of vertices that are not immediately modularizable in the original form of the FT.
2. The modular decomposition of the FT starts simultaneously from all its non-connected primary operators.
3. The primary operators having an operator of the same type as predecessor are merged with their predecessors by transferring all their inputs to the predecessor.
4. The primary operators having a predecessor of a different type are modularized. Those not having repeated event or sub-module as input are temporarily transformed into modules. If not, the set of repeated events of the gate in question will be complete and in that case, a modular representation of the minimal cut sets for their composition could be carried out.
5. The operators already transformed into proper modules or into temporary sub-modules are attached to their predecessors as new components.
6. The stages #3–#5 are repeated until the top event is attained.

End.

It should be noted that the original algorithm [OLM 78] provides for the treatment of the k-out-of-n operators. Another algorithm proposed by Willie is presented in [WIL 78].

▷ **Example 7.3.** The application of the algorithm to the FT of Figure 7.1 will directly lead to the FT of Figure 7.2.

▷ **Example 7.4.** We make use of the preceding algorithm for modularizing the FT, given in Figure 7.3. In this FT, we have two repeated events $\{4, 5\}$

and three primary operators $\{C, E, F\}$; the operator E is reduced to a module and the operators C and F become connected. Figure 7.4 shows an intermediate stage of the algorithm and Figure 7.5 presents the modularized FT.

Modules of the structure function

It will be of interest to be in a position to modularize the structure function of an FT, when it is known that the calculation of the probability of the top event will be based on this function. This brings about a considerable reduction in the complexity of the calculations. The method proposed in [WIL 85] is as follows:

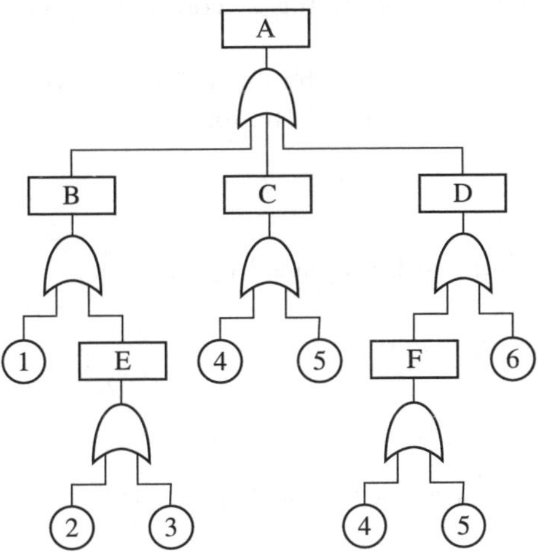

Figure 7.3 *Fault tree to be modularized*

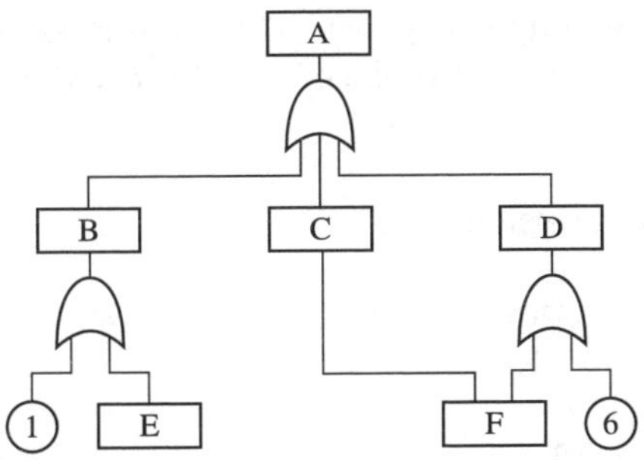

Figure 7.4 *Intermediate fault tree*

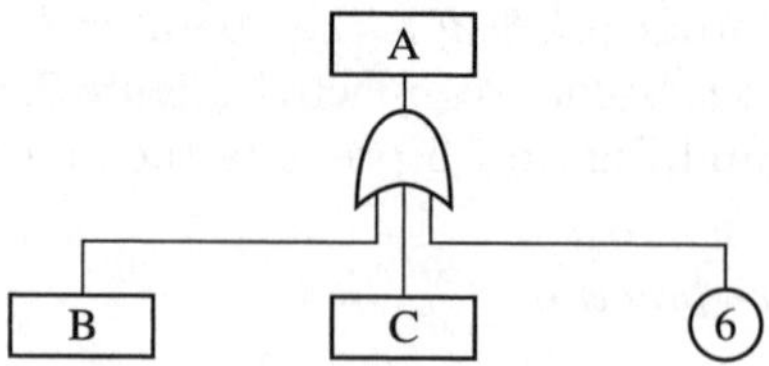

Figure 7.5 *Modularized fault tree*

For a given Boolean function φ, we will examine the couples of the variables (x_i, x_j), for $i \neq j$. Let us assume that such a couple fulfills the two conditions as follows:

(C1) any monomial of φ containing x_i contains x_j as well;

(C2) for any monomial of φ of the form $\overline{x}_i P$ (where P contains neither x_i nor x_j), there is also a monomial of the form $\overline{x}_j P$.

Then:

$x_i x_j$ is a module and can be replaced by y_i

$\overline{x}_i P$ is replaced by $\overline{y}_i P$, and

$\overline{x}_j P$ is eliminated.

Then, y_i can be compared with another x_i, etc.

▷ **Example 7.5.** Let an FT have the structure function as follows:

$$\varphi(\mathbf{x}) = \overline{x}_1 x_3 + \overline{x}_2 x_3 + x_4 x_6 + x_5 + x_1 x_2.$$

Let us examine the couple (x_1, x_2). This couple satisfied conditions $(C1)$ and $(C2)$; consequently, we will replace $x_1 x_2$ by y_1, $\overline{x}_1 x_3$ by $\overline{y}_1 x_3$ and we will eliminate $\overline{x}_2 x_3$.

This leads to the expression:

$$\varphi(\mathbf{x}) = y_1 + \overline{x}_3 + x_5 + x_4 x_6.$$

7.3 Multiphase fault trees

Multiphase systems have been studied by Easary and Ziehms [ESA 75].

7.3.1 *Example*

Let us consider a hydraulic system made up of two pumps, which work independently of each other. This system should function in two phases. In phase 1, the yield required is $d = 50$ units of volume (u.v) per unit time (u.t.); in phase 2, $d = 100$ u.v./u.t.

Given the fact that the flows of the pumps are equal at 50 u.v./u.t./pump, it is evident that in phase 1 one of the two pumps should be functioning, and in phase 2, the two pumps should function at the same time.

The success of the mission implies the success of phase 1 and the success of phase 2. The FT concerning the failure of the mission is given in Figure 7.6.

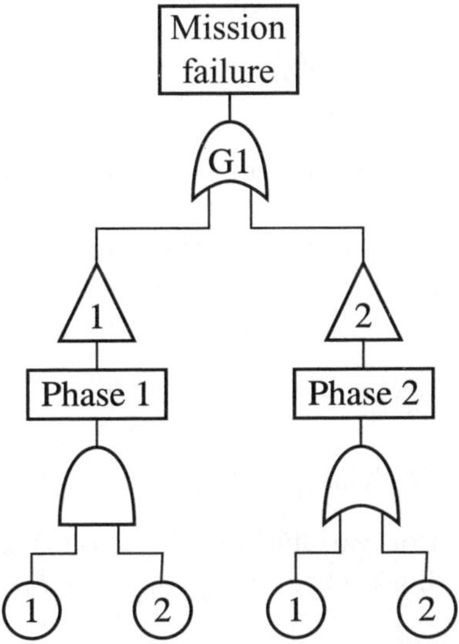

Figure 7.6 *Multiphase fault tree*

For each phase, we will construct an FT and will compile them together by an OR operator.

Let p_{i1} be the probability that the component i will function during phase 1, and p_{i2}, the conditional probability that the component i will function during phase 2, knowing that it functioned during phase 1.

The reliability of the system will be:

– for the phase 1: $p_1 = p_{11} + p_{21} - p_{11}p_{21}$

– for the phase 2: $p_2 = p_{12}p_{22}$,

and for the system in the two phases:

$$r(\mathbf{p}) = p_1 p_2 = (p_{11} + p_{21} - p_{11}p_{21})p_{12}p_{22}.$$

7.3.2 *Transformation of a multiphase system*

Let us consider a system that has to function in m phases $(m > 1)$. Let u_i, $i = 1, ..., m$ be the durations of phases and $u_1 + ... + u_m = \theta$, where θ is the duration of mission of the system. Set $t_j = u_1 + ... + u_j$, $j = 1, ..., m$. We will agree that the system is located in the phase j $(1 \leq j \leq m)$ in the time interval $[t_{j-1}, t_j)$.

For a component k and a fixed time period t, let us consider $X_k(t)$ the random variable with value $\{0, 1\}$ (1: good functioning and 0: breakdown).

The transformation can be carried out as follows:

(a) Replace, in the configuration of the phase J, the component k by a series system with the components: independent $k_1, ..., k_J$, corresponding to the phases $1, ..., J$, whose probabilities of working are:

$$P(Y_{k_1} = 1) = P(X_k(t_1) = 1),$$

$$P(Y_{k_i} = 1) = P(X_k(t_i) = 1 | X_k(t_{i-1}) = 1), \quad i = 2, ..., J.$$

where, Y_{k_j} is the indicator variable for the good working of the imaginary component j, corresponding to the real component k.

(b) Consider that the transformed configuration of a phase is a sub-system operating in series.

7.3.3 *Method of eliminating the minimal cut sets*

A minimal cut set of a phase can be eliminated if it contains a minimal cut set of a later phase.

▷ **Example 7.6.**

– Minimal cut set of the phase 1: $K_{11} = \{1, 2\}$.

– Minimal cut set of the phase 2: $K_{21} = \{1\}$ and $K_{22} = \{2\}$.

The minimal cut sets K_{11} should be eliminated as it contains the minimal cut set K_{21}.

Apart from the phases considered from the working point of view, we can consider the phases concerning the resources of the system, in general, of random duration.

For illustrating this idea, let us consider a non-repairable system of the order n having active parallel structure. When k $(1 \leq k \leq n)$ components are working, the failure rate of each component is λ/k $(\lambda > 0)$. This system, which starts to work at the instant $t = 0$, presents n phases of functioning for random periods. In the phase i, having a duration T_i, there are i components that are working and $n - i$ components that are under breakdown. We can easily see that all the random variables T_i follow the exponential laws of parameter λ. The life duration of the system T, which we make out in n phases, follows a Gamma distribution $\gamma(n, \lambda)$ (see Exercise 2.4).

7.4 Common mode failures

One of the most prominent hypotheses about the reliability of the systems, and more particularly concerning the FTs, is the statistical independence of the basic events.

Two events A and B are said to be statistically independent or independent in probability or simply independent if:

$$P(A \cap B) = P(A)P(B).$$

For three events: A, B and C, we should have:

$$P(A \cap B) = P(A)P(B), \quad P(B \cap C) = P(B)P(C), \quad P(C \cap A) = P(C)P(A),$$

$$P(A \cap B \cap C) = P(A)P(B)P(C).$$

For n events, we should verify $2^n - n - 1$ equalities at the maximum for establishing the independence of these events.

The dependences considered above are concerned with the probability values of the events and are connected to our state of knowledge. Another type of dependence is concerned with the functional dependences connected to the physical world. The functional dependences concerning the failures of the components in a system are called "common modes of failure" (CM).

The common modes of failure (or "the common causes" or simply "the common mode") are simultaneous failures or cascade failures of many components or sub-systems due to a common cause.

The distinction between the functional dependences and dependences linked to the state of our knowledge is not always quite clear. For example, two components produced by the same manufacturer have a tendency to be correlated; this correlation is due to the functional dependences by virtue of a common manufacturing procedure. It cannot be considered a CM, as in general we do not have simultaneous failures or cascade failures. Nevertheless, this distinction, commonly made in the studies of operational safety, turns out to be effective.

A common mode can be on account of a particular error or cause (see the classification of CM in [PAG 80]):

– Design (i.e., error of design).

– Manufacture (i.e., inadequate standards).

– Procedure (i.e., human errors).

– Environment (i.e., vibration, earthquake).

▷ **Example 7.7.** Let us consider a 2-component parallel system with a common electric supply. The failure in the electric supply will constitute a CM and makes both the components out of service at the same time and hence brings about the failure of the system. A first FT, Figure 7.7(a), is constructed without analyzing the CM. This FT leads to a false analysis and probabilistic assessment. If q_e is the failure probability of the electric supply and q_i, $i = 1, 2$ is the failure probability of the component i, we will obtain through FT (a) the failure probability of the system: $P'_S = q_e^2 + q_e(q_1 + q_2) + q_1 q_2$. From the FT (b), where we have taken into account the CM, we will have: $P_S = q_e - q_e q_1 q_2 + q_1 q_2$, which represents the real value. We have: $P_S < P'_S$.

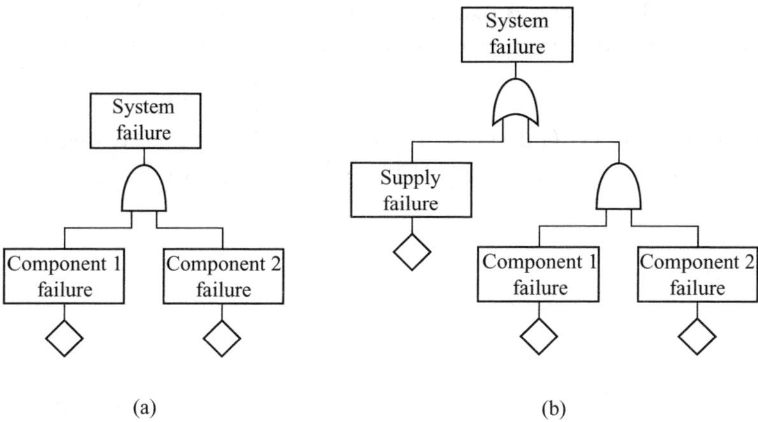

Figure 7.7 *Illustration of separable CM: (a), FT with a common mode, (b) FT with the separate common mode*

There are two methods for dealing with the CM's within the framework of the FTs:

– When the common modes are separable; in this case, we will consider the CM separately, as in the example shown in Figure 7.7.

– When the CMs can be treated by a stochastic process. For example, we can treat certain forms of dependences through Markov chains and then integrate the results in the FT, where the dependent components will be considered as a macro-component.

The method of separable causes consists of considering, in a form that is more general than that illustrated in the FT described above, the distinct causes affecting the sub-sets of a set of components. For example, for a set of three components $\{A, B, C\}$, we will be considering distinct causes affecting the sub-sets of the components $\{A\}, \{B\}, \{C\}, \{A, B\}, \{B, C\}, \{C, A\}$ and $\{A, B, C\}$. For example, the cause affecting the sub-set $\{C\}$ is distinct from the cause affecting the sub-set $\{C, A\}$.

If the probabilities of causes are q_1 for the individual causes, q_2 for the causes of the couples and q_3 for the cause of common failure of three components, then the probability that a component be faulty is: $q_1 + 2q_2 + q_3$.

The probability of failure of a series sub-system having as components A, B and C is equal to $P(C_A \cup C_B \cup C_{AB} \cup C_{BC} \cup C_{CA} \cup C_{ABC})$, (where $C_{(\cdot)}$ designates the causal event of $(\cdot)$). We can develop this probability by the methods studied previously, for example the inclusion-exclusion development.

More generally, we can consider that the instances attributed for different causes described above are governed by distinct stochastic processes.

There are also statistical methods for the study of CMs, but an illustration of these methods is beyond the scope of this book. The interested reader can consult the following references: [CHA 86], [APO 87], [MOS 91].

Chapter 8

Extensions: Non-Coherent, Delay and Multistate Fault Trees

8.1 Non-coherent fault trees

8.1.1 *Introduction*

In the preceding chapters, we studied the coherent fault trees (c-FT), that is, the FTs that can be described by the fundamental operators (OR and AND) and the monoform variables.

This logic is overtaken very fast in the case of the current complex systems, for example: regulation loops, maintenace specifications, etc.

Another family of FT deals with the non-coherent systems, which is called non-coherent FTs (nc-FTs), will form the subject of this chapter. A non-coherent FT, in its restrained form, is described by fundamental operators and contains biform variables. In the case of the nc-FT, we can make use of the NO-operator for representing a complemented variable (see Figure 8.1).

The nc-FTs can also be represented by other operators, such as NOR, NAND. Subsequently, in Figures 8.2, 8.3 and 8.4, we will be giving the operators that are used the most in the case of FTs, and their equivalent representation in terms of AND, OR and NO-operators.

With reference to the above transformations, it often happens that we have successions of NO-operators; we will reduce them as follows:

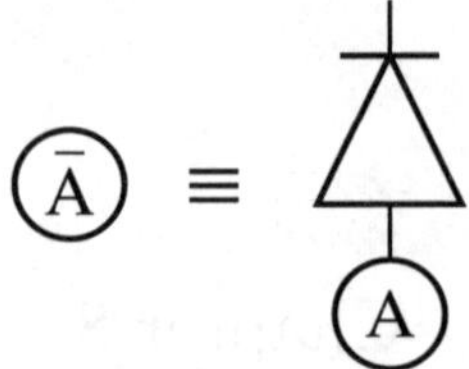

Figure 8.1 *Equivalence of representations of a complemented event*

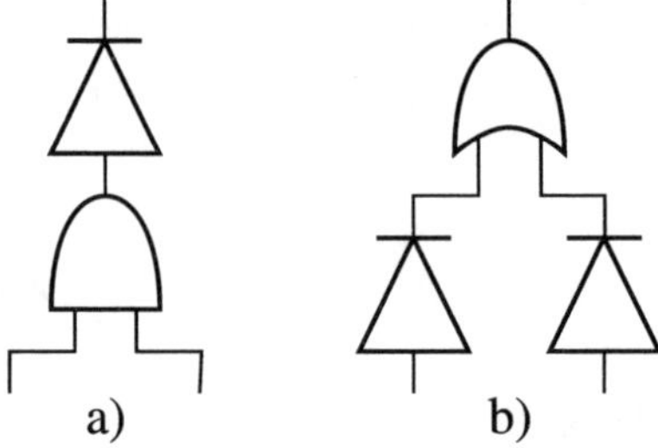

Figure 8.2 *NAND operator (a) and its equivalent transformation (b)*

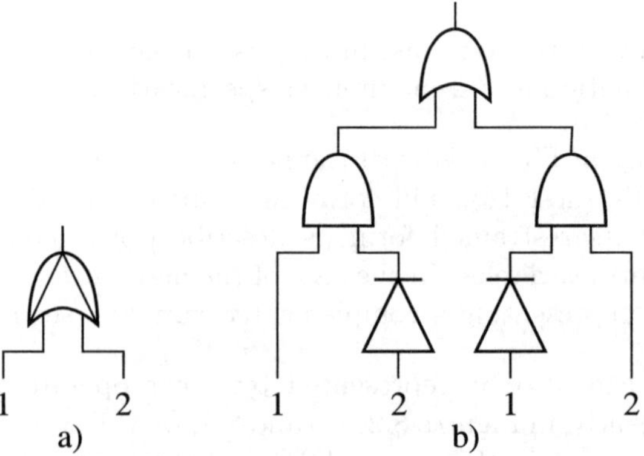

Figure 8.3 *OR-exclusive operator (a) and its equivalent transformation (b)*

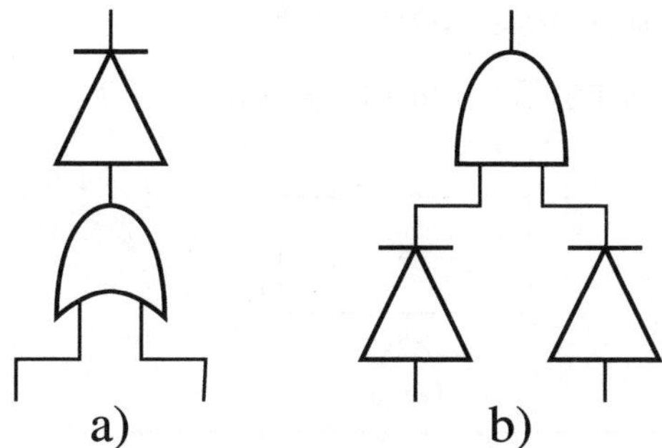

Figure 8.4 *NOR operator (a) and its equivalent transformation (b)*

If the series contains n NO-operators, then we will replace them by a single NO-operator in the case where n is an odd number, and by an identity operator, in the case where n is even. An identity operator is an operator that has no effect on its impact.

We will now consider the nc-FT in their restrained form.

A necessary condition for an FT being non-coherent is that it contains biform variables. This condition is not sufficient. In fact, as we can see in Figure 8.5a, the FT containing the biform variables $\{x, \overline{x}\}$ is a c-FT, because it is equivalent to the c-FT in Figure 8.5b.

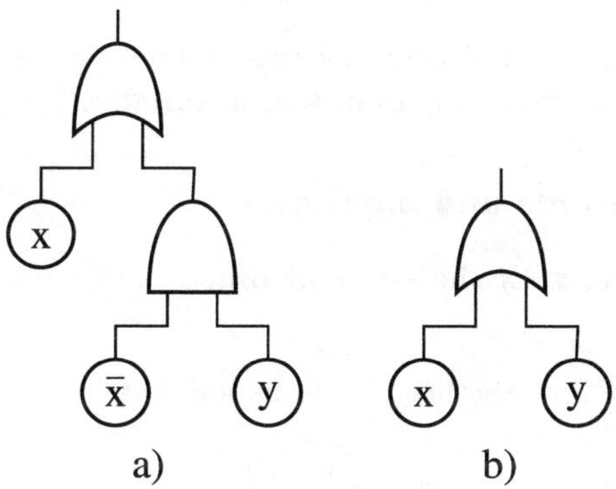

Figure 8.5 *(a) Fault tree containing complementary events (b) Coherent fault tree equivalent to fault tree (a)*

8.1.2 *An example of a non-coherent FT*

Let us consider the FT given in Figure 8.6.

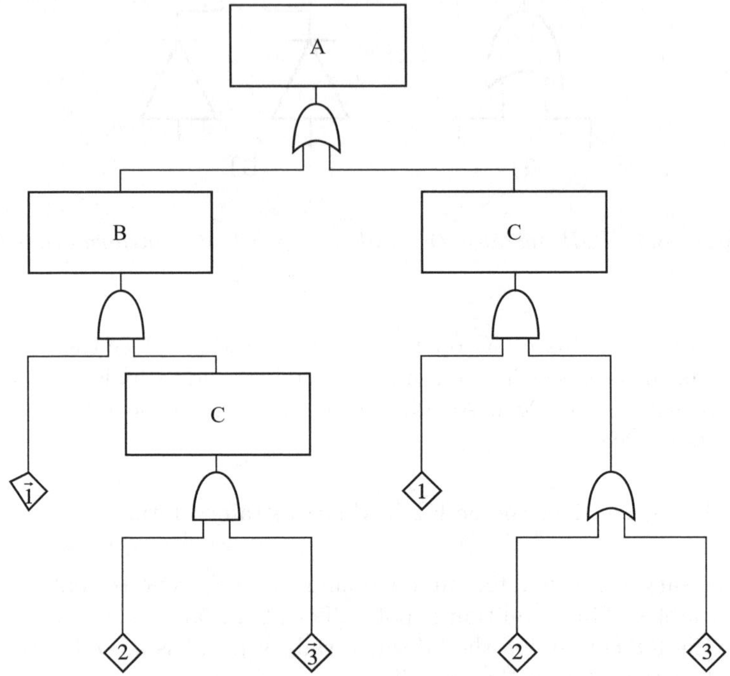

Figure 8.6 *Non-coherent fault tree*

We can see that this FT contains two couples of complementary events: $(1, \overline{1})$ and $(3, \overline{3})$, hence two couples of biform variables: $(x_1, \overline{x}_1)$ and $(x_3, \overline{x}_3)$.

8.1.3 *Prime implicants and implicates*

Before proceeding with the study of prime implicants and implicates, we will define them.

We will define at first the literal, x^a, as follows:

$$x^a = \begin{cases} x & \text{if } a = 1 \\ \overline{x} & \text{if } a = 0. \end{cases}$$

An *implicant* is a conjunction of the no conflicting or repeated literals that imply $\varphi(\mathbf{x})$.

A *prime implicant* is an implicant which is not implied by any other implicant.

An *implicate* is a disjunction of the literals implied by $\varphi(\mathbf{x})$.

A *prime implicate* is an implicate which does not contain any other implicate.

A *base of a function* φ is any disjunction of prime implicant equivalent to φ.

A *complete base* is the disjunction of all the prime implicant.

A *base is irredundant* if it ceases to be a base, when one of the prime implicants in this base is deleted.

A *base is minimal irredundant* when it contains the least number of prime implicants among all the irredundant bases.

A Boolean function can have more than one minimal irredundant base. If the function φ possesses just one base, which is simultaneously complete and irredundant, then φ is coherent.

The same algorithms that are utilized for the c-FT can be employed for the nc-FT, but on considering, in addition, the terms of consensus.

Let us consider the identity:

$$xy + z\overline{x} = xy \dot{+} z\overline{x} \dot{+} yz. \tag{8.1}$$

The term yz is called the *term of consensus.* A particular version of the above identity is given below:

$$x + \overline{x}y = x \dot{+} y.$$

Thus, when we, by employing the algorithms developed for the c-FT, obtain the list of cut sets (implicants), we have to take into account the terms of consensus during the reduction.

▷ **Example 8.1.** We will obtain the prime implicants of the FT in Figure 8.6 On applying the Mocus algorithm (see Chapter 4) we take:

$$I_1 = \overline{x}_1 x_2 \overline{x}_3$$

$$I_2 = x_1 x_2$$

$$I_3 = x_1 x_3$$

I_1 and I_2 give the consensus $I_4 = x_2\overline{x}_3$, which will be a fourth implicant. The implicant I_1 will be eliminated as it is no longer prime. Thus, the prime implicants of the studied FT are: I_2, I_3 and I_4. Here we have the following theorem of Worel *et al.* [WOR 78]: "If the Boolean function φ has a dual function that does not contain zero product, then the prime implicants are obtained without applying the term of consensus".

▷ **Example 8.2.** The following function

$$\varphi(\mathbf{x}) = (x_1 \dotplus x_2 x_3 \overline{x}_4)(\overline{x}_1 \dotplus x_4 \dotplus \overline{x}_3)$$

has a dual function:

$$\varphi^D(\mathbf{x}) = \overline{x}_1(\overline{x}_2 \dotplus \overline{x}_3 \dotplus \overline{x}_4) \dotplus \overline{x}_4, x_1, x_3.$$

without zero product.

As a result, we will obtain the prime implicants of φ through simple development, as well as the reductions of the coherent functions without making use of the term of consensus.

$$\varphi(\mathbf{x}) = x_1 x_4 \dotplus x_1 \overline{x}_3 \dotplus \overline{x}_1 x_2 x_3 \overline{x}_4.$$

8.1.4 *Probabilistic study*

The probabilistic assessment of the nc-FT can be done by the methods that we have developed in the case of the c-FT.

During the inclusion-exclusion development, we have to take into account the reduction due to complementary events (i.e., $A \cap \overline{A} = \emptyset$). Thus, the probability of intersection of two implicant containing at least one complementary event is equal to zero. The terms of the development to be calculated for a given precision are not more important than those for the c-FT.

During the factorization, it will be necessary to integrate the complementary events into the list of repeated events.

The recursive methods are valid also for the nc-FT. It has to be mentioned that, when in a term we have the conjunction of two complementary events, then the term will be eliminated.

The calculation of the bounds is applicable also for the nc-FT, except for the "min-max" bounds [APOS 77].

▷ **Example 8.3.** (Example 8.1 contd.). We get:

$$Q = q_1 q_3 + q_2 + q_1 q_2 - q_1 q_2 q_3 - q_1 q_2 p_3 = q_1 q_3 + q_2 p_3.$$

As for the analysis of uncertainty, the same method used for the coherent FTs can be employed. As regards the factors of importance, we will use the same relationships as for those the coherent fault trees, but wherever we have ΔQ, we need to substitute $|\Delta Q|$, [JAC 83].

8.2 Delay fault trees

8.2.1 *Introduction*

In this section, we will study the fault trees comprising, apart from the OR and AND operators, the DELAY operators, which we call the *delay fault trees.*

The DELAY operator contains a single input and a single output. The output event is realized when that at the input is realized for at least a time period τ, which is equal to the delay given by the operator.

We have developed algorithms for the transformation of a delay fault tree into a standard fault tree.

8.2.2 *Treatment*

Let us consider a DELAY(τ) operator having for input an AND operator, itself having two inputs. We can transform this structure into another equivalent structure: this consists of an AND operator having for input two DELAY(τ) operators (see Figure 8.7a) and b)).

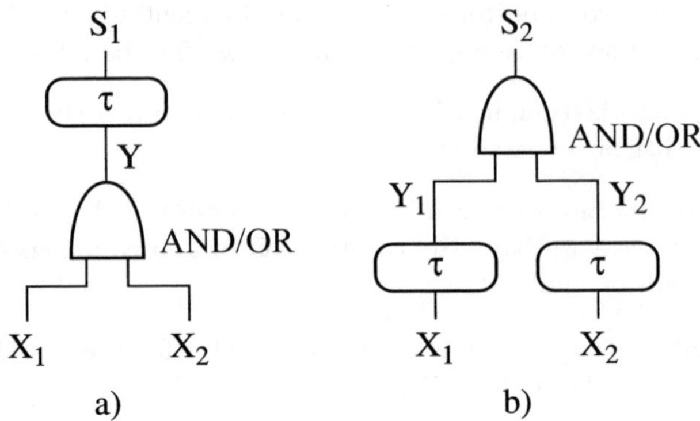

Figure 8.7 *Transformation of a structure comprising a delay operateor as output a) into an equivalent structure with two delay operators as input b)*

We have:

$$S_1(t) = Y(t-\tau) = X_1(t-\tau)X_2(t-\tau)$$

and

$$S_2(t) = Y_1(t)Y_2(t) = X_1(t-\tau)X_2(t-\tau)$$

whence

$$S_1(t) = S_2(t), \quad t \geq 0.$$

Let us now consider a DELAY(τ) operator having for input an OR operator, which in turn has two inputs. We can transform this structure into another equivalent structure: this one consists of an OR operator having for input two DELAY(τ) operators. (see Figure 8.7a) and b)).

We have:

$$S_1(t) = Y(t-\tau) = X_1(t-\tau) \dot{+} X_2(t-\tau)$$

and

$$S_2(t) = Y_1(t) \dot{+} Y_2(t) = X_1(t-\tau) \dot{+} X_2(t-\tau),$$

whence

$$S_1(t) = S_2(t), \quad t \geq 0.$$

We have considered two operators with two inputs for simplifying the presentation. The result is valid for operators having more than two inputs.

Let $G = (X, U)$ be a delay fault tree (containing at least one delay operator), and $D \in X$ be a DELAY(τ) operator. Let $\mathcal{D}(D)$ be the domain of D (the set of the basic events of the sub-fault tree having D as its top operator).

We can transform the fault tree described above into an equivalent fault tree. It suffices to eliminate the operator D by transforming at first all the events of the fault tree belonging to $\mathcal{D}(D)$ into "delay events" (delay components [LIM 90]). The proof of this assertion is immediate, thanks to the two proof of transformation of operators described above.

The treatment of the fault tree thus obtained is done in a traditional manner, that is: obtaining the cut sets, reduction and evaluation. See also the delay distribution in section 1.1.8.

8.3 FTs and multistate systems

In this section, we will give a brief introduction to the multistate systems by presenting the diverse types of these systems, a general definition of the structure function as well as the definition of the reliability in this general context. We will then present the two types of FT for modelizing the multistate systems: FTs with restrictions and multistate FTs.[1]

8.3.1 *Multistate systems*

Up to now we have been studying systems with state spaces $\mathcal{E}_i = \mathcal{E} = \{0, 1\}$, for any component i, that is to say, "binary systems". The binary systems constitute the essential paradigm of the theory of reliability of systems.

The difficulties of the binary model for describing more and more complex systems have led, since the beginning of the 1970s, to considerations that are more ambitious than the simple "reliability of switches". Thus, through the natural route, the negation of the binary character on the state spaces of the components and of the systems led to multistate systems.

Postelnicu introduced in 1970 [POS 70] the state space $[0, 1] \subset \mathbb{R}$, for the components and the systems. For these systems, and without algebraic consideration, he deduced a certain number of probabilistic results. This idea of

1 The reader solely interested in FTs can move on directly to section 8.3.4.

"extension" led the analysts to consider other types of systems; the systems studied in the books are, to a large extent, systems of discrete state spaces. Thus, we can classify them, depending on their state spaces, into three big families:

Systems with totally ordered discrete state spaces

This first family contains three types of systems:

– The multistate systems with $\mathcal{E}_i = \mathcal{E} = \{0, 1, ..., M\}$ (studied by Barlow and Wu [BAR 78], El-Neweihi, Proschan and Sethuraman [PRO 78], Griffith [GRI 80 & 82], Natvig [NAT 82] and Block and Savits [BLO 82]).

– The systems with $\mathcal{E} = \{0, 1, ..., M\}$ and $\mathcal{E}_i = \{0, 1, ..., M_i\}$ (studied by Ohio, Hiroso and Nishida [OHI 84], Janan [JAN 85] and Wood [WOO 85]).

– The systems with $\mathcal{E} = \{0, 1, ..., M\}$ and $\{0, M\} \subset \mathcal{E}_i \subset \mathcal{E}$ (studied by Natvig [NAT 82]).

It should be noted that the second and third types, as we will see later, can be considered particular cases of the first type.

Systems with partly ordered discrete state spaces

The systems of this family are modeled by the definition of binary variables for each mode of failure. Thus:

– Caldarola [CAR 80] and Xizhi [XIZ 84] consider Boolean algebras with "restrictions" to the variables.

– Garriba *et al.* [GAR 80 & 85] consider Boolean algebras without restriction.

Systems with continuous state spaces

– Ross [ROS 79] considers systems with state spaces $\mathbb{R}_+$

– Postelnicu [POS 70], Baxter [BAX 84], and Montero [MON 90] consider state spaces the closed $[0, 1] \subset \mathbb{R}$.

8.3.2 *Structure function*

Let there be a system S, with C the set of its components (if $|C| = n < \infty$, and $n \geq 1$, the system is said to be of order n); for each component $i \in C$, we will assign a function $x_i(t)$, (if $x_i(t) = k$, $k \in \mathcal{E}_i$, the component i is said to be in the state k, at time t); for the set of the components, let us consider the vector $\mathbf{x}(t) = (x_1(t), ..., x_n(t))$. We will define a function φ over $\mathcal{E}_1 \times ... \times \mathcal{E}_n$ and values within $\mathcal{E}$ (state space of the system) called the structure function. The system in question is noted as $S = (C, \times_{i=1}^{n} \mathcal{E}_i, \mathcal{E}, \varphi)$.

More exactly, the structure function is defined as a measurable application. We define the measurable space $(\mathcal{E}_i, \mathcal{B}(\mathcal{E}_i))$, for all $i \in C$, and the product measurable space $(\mathcal{E}_1 \times ... \times \mathcal{E}_n, \mathcal{B}(\mathcal{E}_1 \times ... \times \mathcal{E}_n))$, where $\mathcal{B}(\cdot)$ is the Borelian σ-field generated by the family of opens $(\cdot)$; then, the structure function is a measurable application:

$$\varphi\colon (\mathcal{E}_1 \times ... \times \mathcal{E}_n, \mathcal{B}(\mathcal{E}_1 \times ... \times \mathcal{E}_n)) \longrightarrow (\mathcal{E}, \mathcal{B}(\mathcal{E})).$$

The product set $\mathcal{E}_1 \times ... \times \mathcal{E}_n$, provided with the (product) and $\dot{+}$ (sum), is called a *Boolean lattice* (complemented distributive lattice). The elements of $\mathcal{E}_1 \times ... \times \mathcal{E}_n$ are n-dimensional vectors, noted as $\mathbf{x} = (x_1, ..., x_n)$. The Boolean lattice possesses the usual properties: associativity, commutativity and distributivity. The binary case that we presented in Chapter 2 is a particular example.

As for the binary case, the product set $\mathcal{E}_1 \times ... \times \mathcal{E}_n$ is partly ordered with the relationship "$\leq$", called *inclusion*. In the usual manner, we will also define the strict inclusion, noted "$<$" and the equality, noted "$=$".

The following definitions and properties relate to systems with discrete states.

Now, we use as support, so as to give the basic definitions, the system: $S_M = (C, \mathcal{E}^n, \mathcal{E}, \varphi)$, with $\mathcal{E} = \{0, 1, ..., M\}$. In fact, this type of systems also represents the systems with $\mathcal{E}_i = \{0, 1, ..., M_i\}$ and $\mathcal{E} = \{0, 1, ..., M_0\}$, because, in the latter case, we can define $M = \max\{M_1, ..., M_n, M_0\}$ and consider the system S_M with, for structure function, the function $\widehat{\varphi} : \mathcal{E}^n \to \mathcal{E}$, with $\widehat{\varphi}(\mathbf{x}) = \varphi(x_1 \dot{+} M, ..., x_n \dot{+} M)$.

Dual structure function: for a given structure function φ, we define its dual function φ^D through the relationship:

$$\varphi^D(\mathbf{x}) = M - \varphi(\mathbf{M} - \mathbf{x}).$$

Essential component: a component $i \in C$ is said to be essential or relevant for the structure function φ:

If there exists an $\mathbf{x} \in \mathcal{E}^n$ such that $\varphi(M_i, \mathbf{x}) \neq \varphi(0_i, \mathbf{x})$, with $(a_i, \mathbf{x}) = (x_1, ..., x_{i-1}, a, x_{i+1}, ..., x_n)$.

The component $i \in C$ is said to be *strictly essential* for the structure function φ if:

For all $z, y \in \mathcal{E}, y \neq z$, there exists an $\mathbf{x} \in \mathcal{E}^n$ such that $\varphi(z_i, \mathbf{x}) \neq \varphi(y_i, \mathbf{x})$.

Monotone system: the system S is said to be monotone if:

– φ is non-decreasing with respect to $\mathbf{x} \in \mathcal{E}^n$;

– for all $\mathbf{x} \in \mathcal{E}^n$, we have

$$\min_{i \in c}\{x_i\} \leq \varphi(\mathbf{x}) \leq \max_{i \in c}\{x_i\}.$$

Coherent component and *coherent system*: the component $i \in C$ is said to be strictly coherent for the structure function φ if:

$$\text{For all } j, k \in \mathcal{E}, \text{ there exists an } \mathbf{x} \in \mathcal{E}^n$$
$$\text{such that } \varphi(k_i, \mathbf{x}) = j \text{ if and only if } k = j.$$

When all the components $i \in C$ are strictly coherent, the system is said to be strictly coherent.

Elementary systems: the three types of systems given below are called elementary systems:

– Series system at the level m if

$$\varphi(\mathbf{x}) = m \Leftrightarrow \min_{i \in C} x_i = m.$$

– Parallel system at the level m if

$$\varphi(\mathbf{x}) = m \Leftrightarrow \max_{i \in C} x_i = m.$$

– System k-out-of-n at the level m if

$$\varphi(\mathbf{x}) = m \Leftrightarrow \max_{i \in C}\left\{i : \sum_{j=1}^{n} \mathbf{1}_{\{x_j \geq i\}} \geq k\right\} = m.$$

In fact, the "series" systems and the systems "in parallel" are particular cases of the systems "k-out-of-n". A "series" system is a system "n-out-of-n" and a system "in parallel" is a system "1-out-of-n".

Any system that can be composed uniquely of elementary sub-systems is called a *system of elementary structure.* In the opposite case, it is called a system of *complex structure.*

Minimal sets: the minimal sets (minimal cut sets and paths) play a preponderant role in the study of systems with complex structure; they allow us to construct their structure function.

A vector $\mathbf{x}$ is a *path at level* m if $\varphi(\mathbf{x}) \geq m$; it is a *path vector at minimal level* m if $\varphi(\mathbf{x}) = m$; it is a *minumal path vector at maximal level* m if $\varphi(\mathbf{y}) < m$, for $\mathbf{y} < \mathbf{x}$.

A vector $\mathbf{x}$ is a *cut set at level* m if $\varphi(\mathbf{x}) < m$; it is a cut set vector *at minimal level* m if $\varphi(\mathbf{x}) = m - 1$; it is a *minimal cut set vector at minimal level* m if $\varphi(\mathbf{y}) \geq m$, for $\mathbf{y} \geq \mathbf{x}$.

Module: (A, χ) is a module of the system $S_M = (C, \mathcal{E}^n, \mathcal{E}, \varphi)$ if $A \subset C$ and $\varphi(\mathbf{x}) = \psi(\chi(\mathbf{x}^A), \mathbf{x}^{\overline{A}})$, where ψ is a coherent structure function called an *organizing function,* and $\mathbf{x}^A$ is the restriction of the vector $\mathbf{x}$ over A.

The set of disjoint modules $\{(A_1, \chi_1), ..., (A_r, \chi_r)\}$ is a modular decomposition of the system $S_M = (C, \mathcal{E}^n, \mathcal{E}, \varphi)$, with, as organizing function, the function ψ, if:

(i) $C = A_1 \cup ... \cup A_r$

(ii) $\varphi(\mathbf{x}) = \psi(\chi_1(\mathbf{x}^{A_1}), ..., \chi_r(\mathbf{x}^{A_r}))$.

8.3.3 *Stochastic description and function of reliability*

Let a multistate system of order n be $S = (C, \mathcal{E}^n, \mathcal{E}, \varphi)$. The performance process of the component $i \in C$ is the right continuous stochastic process $\{X_i(t); t \in \mathcal{T}\}$, defined over a probability space $(\Omega_i, \mathcal{A}_i, P_i)$ and with values in $\mathcal{E}$. The set of the time is: $\mathcal{T} = \mathbb{R}_+$ or $\mathcal{T} = \mathbb{N}$.

The simultaneous performance process is the vector stochastic process $X(t) = (X_1(t), ..., X_n(t); t \in \mathcal{T})$ over the space of product probability: $(\Omega, \mathcal{A}, P) = (\times_C \Omega_i; \times_C \mathcal{A}_i; \times_C P_i)$ and with values in $\mathcal{E}^n$.

The performance process of the system is the right-handed continuous stochastic process $\{\varphi(X(t)); t \in \mathcal{T}\}$, defined over $(\Omega, \mathcal{A}, P)$ and with values in $\mathcal{E}$.

In line with the above considerations, we can define the reliability function as follows:

– Reliability of the system at the level z at time t, $R(z,t)$:

$$R(z,t) = P(\varphi(X(u)) \geq z, \quad \forall u \in \mathcal{T}(t)),$$

where $\mathcal{T}(t) = \mathcal{T} \cap [0,t]$.

– Instantaneous availability at the level z at time t, $A(z,t)$:

$$A(z,t) = P(\varphi(X(t)) \geq z),$$

– Maintainability of the system at the level z at time t, $M(z,t)$:

$$M(z,t) = 1 - P(\varphi(X(u)) < z, \quad \forall u \in \mathcal{T}(t)).$$

▷ **Example 8.4.** In the case of binary systems, we can find again the well-known formulae:

$$R(t) = P(\varphi(X(u)) = 1, \forall u \in \mathcal{T}(t)),$$

$$A(t) = P(\varphi(X(t)) = 1),$$

$$M(t) = 1 - P(\varphi(X(u)) = 0, \forall u \in \mathcal{T}(t)).$$

In the discussion that follows, we have two types of fault trees for studying the multistate systems with discrete state space: the fault trees with restrictions (FT-r) and the multistate fault trees (m-FT).

8.3.4 *Fault trees with restrictions*

The FT-r within the framework of the multistate systems are similar to the binary FTs, except for the fact that many basic events are mutually exclusive, because they are defined over the state space of a same component of the system, which in general exhibits multistate. From this point of view, the FT-r are a generalization of the nc-FT. It should be noted that in the case of the FT-r, the mutually exclusive basic events are not necessarily complementary as in the case of the nc-FT.

The treatment of the FT-r necessitates considering these dependences at the level of the algebraic analysis. Caldarola [CAR 80] has proposed a Boolean algebra with restrictions for variables.

Let there be a component i of the system and $\mathcal{E}_i = \{0, 1, ..., n_i\}$ be its state space, and the variable x_i with values $\mathcal{E}_i$. Let us define the binary variables as follows:

$$x_{ij} = \mathbf{1}_{\{x_i = j\}}, \quad i \in C, j \in \mathcal{E}_i.$$

It is clear that we have:

$$x_{i0} \dot{+} x_{i1} \dot{+} ... \dot{+} x_{in_i} = 1, \quad i \in C, \tag{8.2}$$

and

$$x_{ij} x_{ik} = 0; \quad j \neq k. \tag{8.3}$$

The Boolean algebra with the restrictions proposed in [CAR 80] is concerned with the two relationships (8.2) and (8.3).

We also define the complementary variable:

$$\overline{x}_{is} = x_{i0} \dot{+} x_{i1} \dot{+} ... \dot{+} x_{i,s-1} \dot{+} x_{i,s+1} \dot{+} ... \dot{+} x_{in_i}, \quad s \in \mathcal{E}_i. \tag{8.4}$$

It is important to note that within the framework of the Boolean algebra with restrictions, the state spaces are not ordered. This translates in a better manner the reality of the system concerning the reliability. As underlined by Caldarola, an electric switch is not more faulty under the "opening failure" mode than that under the "closing failure" mode.

For obtaining the prime implicants of an FT-r, we will obtain the implicants of the top event (for example, by using the Mocus algorithm), and then, from the Nelson algorithm [NEL 54 & 55], we will obtain an irredundant base for the Boolean function of the top event.

▷ **Example 8.5.** Let us consider the structure function of an FT obtained from a Mocus-type algorithm. The four components of the system implied in this FT have state spaces of cardinalities 4, 2, 3 and 3. The binary variables defined for this problem are as follows:

Component 1: x_{10}, x_{11}, x_{12}, x_{13}.
Component 2: x_{20}, x_{21}.
Component 3: x_{30}, x_{31}, x_{32}.
Component 4: x_{40}, x_{41}, x_{42}.

The events $\{X_{13} = 1\}$, $\{X_{21} = 1\}$, $\{X_{32} = 1\}$ and $\{X_{42} = 1\}$ designate, respectively, the perfect states of the four components.

The structure function of the FT, obtained by a Mocus-type algorithm, is given below:

$$\varphi(\mathbf{x}) = x_{21} \dot{+} x_{32}x_{11} \dot{+} x_{32}x_{12} \dot{+} x_{42}x_{13} \dot{+} x_{12}x_{33}x_{41} \dot{+} x_{31}x_{11}x_{43}$$
$$\dot{+} x_{31}x_{13}x_{43} \dot{+} x_{31}x_{41} \dot{+} x_{11}x_{33}x_{43},$$

where $\mathbf{x} = (x_{10}, x_{11}, x_{12}, x_{13}, x_{20}, x_{21}, x_{30}, x_{31}, x_{32}, x_{40}, x_{41}, x_{42})$.

On applying the Nelson algorithm [NEL 54 & 55], we will obtain:

$$\varphi(\mathbf{x}) = x_{21} \dot{+} x_{32}x_{12} \dot{+} x_{42}x_{13} \dot{+} x_{12}x_{41} \dot{+} x_{31}x_{13} \dot{+} x_{31}x_{41} \dot{+} x_{11}.$$

8.3.5 *Multistate fault trees*

Multistate fault trees are the trees whose "basic events" do not represent events but components; the "intermediate events" represent the sub-systems; the "top event" represents the system. The operators are generalized, and they represent the applications of the input spaces to the output space (the spaces are discrete).

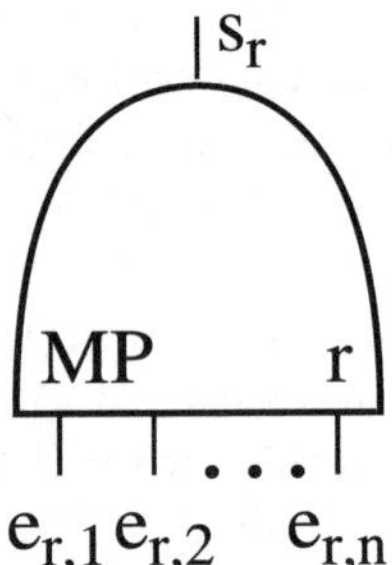

Figure 8.8 *Multistate operator with n inputs*

The operator in Figure 8.8 represents the following application:

$$f : \mathcal{E}_{r,1} \times ... \times \mathcal{E}_{r,n} \to \mathcal{E}_r,$$

where $\mathcal{E}_{r,i}$, $(i = 1, .., n)$ is the space of the i-th input, and $\mathcal{E}_r$ is the output space.

Given the high complexity concerning the treatment of the m-FT, we limit ourselves in pratice to the operators having no more than two inputs. An m-FT can be represented by k FT-r, where $k =$ card $\mathcal{E}$, $\mathcal{E}$ being the state space of the top.

▷ **Example 8.6.** Let us reconsider the example already illustrated in Chapter 7, concerning the two hydraulic pumps. Let us consider three states for the pumps, and three states for the system, that is to say: $\mathcal{E}_1 = \mathcal{E}_2 = \mathcal{E} = \{0, 1, 2\}$. For the pumps we have 2: perfect state and output equal to 50; 1: partial failure and output equal to 25; 0: complete failure and output equal to 0. For the system, we have 2: perfect state and output between 75 and 100; 1: partial failure and output between 25 and 75; 0: complete failure and output less than 25.

The multistate operator describing the state of the system as a function of the states of its components is given in Figure 8.9. The matrix translates the application f given above.

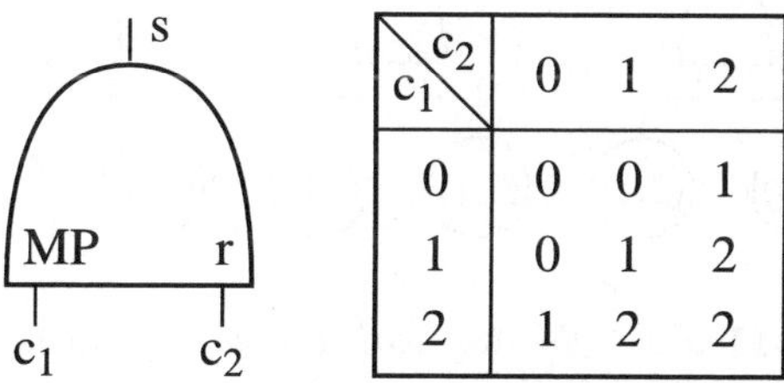

c_1 \ c_2	0	1	2
0	0	0	1
1	0	1	2
2	1	2	2

Figure 8.9 *Multistate operators with two inputs*

The operator (FT-mp) given in Figure 8.9 can be represented by the two FT-r given in Figures 8.10 and 8.11.

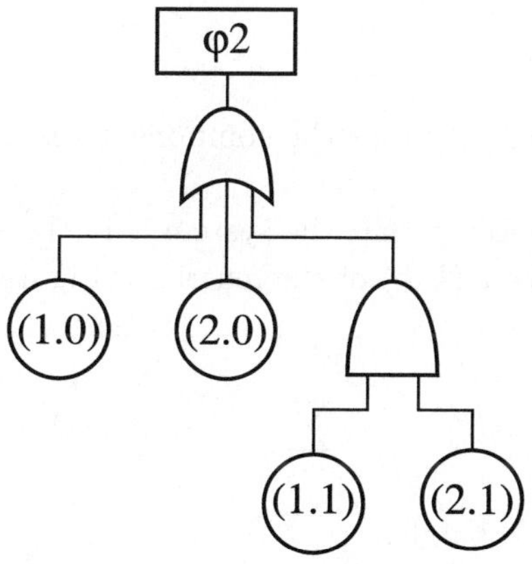

Figure 8.10 *FT-r concerning the state 2 of the system*

The notation (i, j) concerning the basic events of the FTs means the occurrence of the state j of the component i. φ^k designates the non-occurrence of the state k of the system, that is, the level of performance of the system will be less than k. Thus, the FT in Figure 8.11 represents the top event, "not the nominal regime". The FT in Figure 8.10 represents the "non-occurrence of the partial failure of the system", that is, the complete failure.

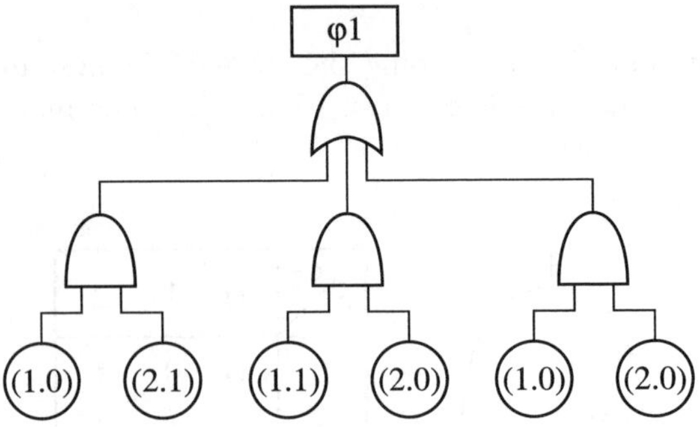

Figure 8.11 *FT-r dealing with the state 1 of the system*

The operators in the FTs described above are binary operators AND and OR. We could have also constructed the FTs dealing with the different states of the system without considering the order relationships in the state spaces.

Concerning the probability of not obtaining the level of performance j, we can assess it either directly through the m-FT of Figure 8.9, or through the FT-r described above.

Let us note:

- $p_{ij}(t)$ the probability that the component i is in the state j at the times t,
- $p_j(t)$ the probability that the system is in the state j at the time t,
- $S(k, h)$ the element (k, h) of the matrix of Figure 8.9.

Then, we have:

$$p_j(t) = \sum_{(k,h):S(k,h)=j} p_{1h}(t)p_{2k}(t).$$

For example, the probability that the performance of the system be less than 2, at time t, is:

$$p_0(t) + p_1(t) = p_{10}(t)p_{20}(t) + p_{11}(t)p_{20}(t) + p_{10}(t)p_{21}(t). \tag{8.5}$$

From the m-FT in Figure 8.9, we have:

$$\varphi^2 = x_{10} \dot{+} x_{20} \dot{+} x_{11}x_{21}.$$

whence we obtain again the expression (8.5). It should be noted that: $x_{10}x_{11} = 0$ and $x_{20}x_{21} = 0$.

It should also be noted that the probabilities $p_{ij}(t)$ as shown above are obtained generally from the resolution of the stochastic processes describing the behavior of the components in the system.

Chapter 9

Binary Decision Diagrams

9.1 Introduction

In recent years, new algorithms concerning fault trees was proposed by Coudert and Madre [COU 92], [COU 94] designated under the term of binary decision diagrams (BDD). It is based on a kind of reduction of the factorization tree of Shannon, as tuned in the treatments of Boolean functions by Akers [AKE 78], and later by Bryant [BRY 87], [BRY 92]. With the aid of this new technique, a complete algorithmics (probabilistic assessment, logical analysis and calculation of the factors of importance) has already been proposed in the literatures [RAU 93], [ODE 95], [ODE 96].

The algorithms based on the BDD are exceptionally rapid. They have made possible the treatment of difficult test cases in a very short space of time (some seconds) instead of very long time periods (some hours, even some days) even with the most efficient traditional algorithms.

Nevertheless, it has to be noted that these algorithms have not altered the nature of the problem, which still continues to be NP-complete. As a result, there can be cases that are difficult to resolve, even with these algorithms.

9.2 Reduction of the Shannon tree

9.2.1 *Graphical representation of a BDD*

Beginning from the tree developed by Shannon, we will proceed with its reduction starting, in principle, from the bottom towards the top, while eliminating the vertices that are not useful. We eliminate a vertex in the following two cases:

– When its two children head for the same vertex, or

– When it is equivalent to another vertex.

Two vertices are said to be equivalent if they carry the same variable and if their children (1) and (0) head for two equivalent vertices, respectively. This definition is applied recursively from the bottom towards the top of the tree. When the two children, (1) and (0), of a vertex (*b*) head for the same vertex (*c*), (Figure 9.1(a)), then vertex (*b*) will be eliminated and its inputs will be directed towards vertex (*c*) (Figure 9.1(b)).

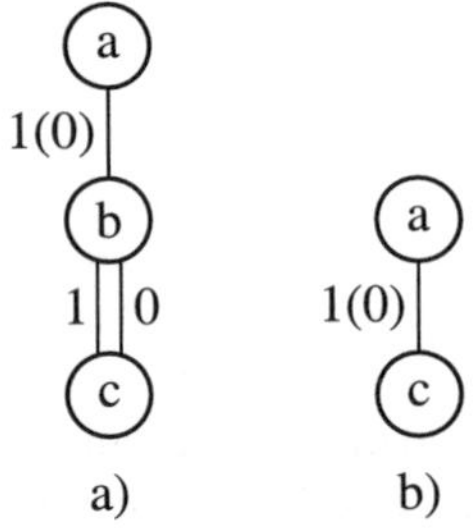

Figure 9.1 *Eliminating vertex (b) in (a)*

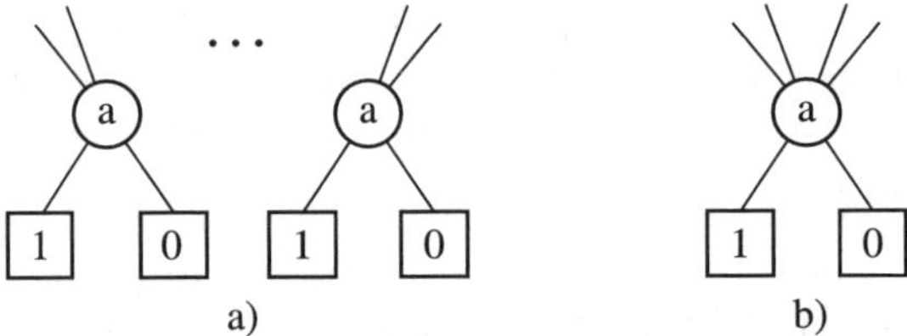

Figure 9.2 *Eliminating one of the two equivalent vertices (a)*

When two or more vertices are equivalent, we retain one of them and eliminate the others by directing their inputs to the retained vertex.

In Figure 9.2(a), the two vertices (a) are equivalent. We will eliminate one of them and will direct its inputs towards the other (Figure 9.2(b)).

▷ **Example 9.1.** With a view to giving a good illustration of the reduction, we will give an example for obtaining the reduced Shannon tree for the following Boolean function:

$$\varphi(x_1,x_2,x_3) = x_1x_2 \dot{+} x_3.$$

Figures 9.3(a), (b) and (c) illustrate the stages of reduction. Figure 9.3(a) shows the developed Shannon tree. It has to be noted that the vertices of the variables are represented by circles, whereas the vertices of values 1 and 0 are represented by rectangles. Figure 9.3(b) represents an intermediate stage of reduction, and Figure 9.3(c) represents the reduced Shannon tree or the BDD.

The Shannon tree (developed or reduced) is not unique; it depends on the order of variables. However, each developed tree has a corresponding unique reduced tree, that is, when we fix the order of the variables, the BDD that we obtain is unique. In this case, we will discuss the ordered BDD or OBDD [BRY 92].

In the preceding example, the order of variables was: $x_1 < x_2 < x_3$. On changing the order, we will obtain a different BDD. For example, by adopting the order: $x_3 < x_1 < x_2$, we will obtain the BDD in Figure 9.4.

Nevertheless, all the BDDs obtained by adopting different orders of variables are equivalent. The expressions of Boolean functions obtained through the two BDDs represented in Figures 9.3(c) and 9.4 are, respectively (we obtain them by enumerating the paths of the root at the vertex 1, in the respective BDDs):

$$\varphi_1(x_1,x_2,x_3) = x_1x_2 + x_1\overline{x}_2x_3 + \overline{x}_1x_3,$$

$$\varphi_2(x_1,x_2,x_3) = x_3 + x_1x_2\overline{x}_3$$

We can easily verify that the above two functions φ_1 and φ_2 are equivalent to φ.

9.2.2 *Formal BDD*

The BDD can also be obtained in a formal manner without using graph representatives. We will be presenting such formalism in the following discussion.

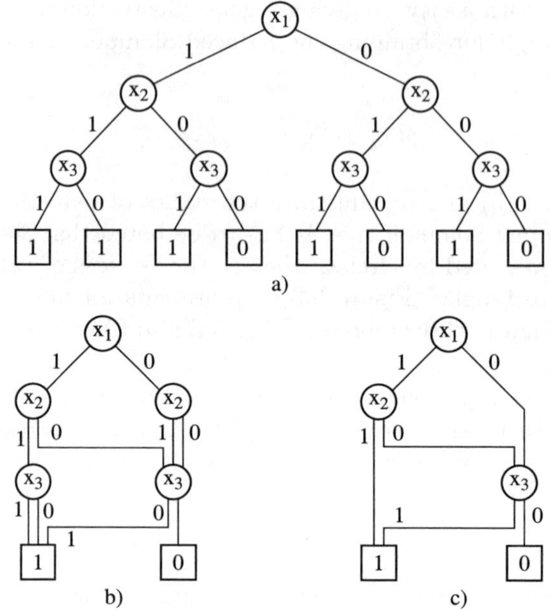

Figure 9.3 *Reduction of the Shannon tree (BDD)*

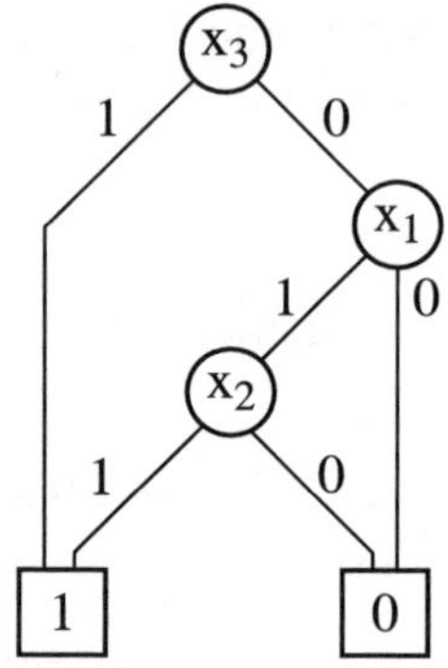

Figure 9.4 *BDD according to the order $x_3 < x_1 < x_2$ of variables*

Let the structure function be φ, and the factorization be as follows.

$$\varphi(\mathbf{x}) = x_i \varphi(1_i, \mathbf{x}) \dot{+} \overline{x}_i \varphi(0_i, \mathbf{x}).$$

We will represent this operation through an ordered triple,

$$< x_i, \varphi(1_i, \mathbf{x}), \varphi(0_i, \mathbf{x}) >,$$

also called *if-then-else*. In a general manner, for the three given Boolean functions f, g and h, we will designate the following operation:

$$< f, g, h >= fg \dot{+} \overline{f}h.$$

The usual operations between the Boolean functions can be carried out with the help of this operation. For example:

$$\begin{aligned} f &=< f, 1, 0 > \\ fg &=< f, g, 0 > \\ f \dot{+} g &=< f, 1, g > \\ f \oplus g &=< f, \overline{g}, g > \end{aligned}$$

Let us consider the Boolean function $\varphi(\mathbf{x})$ designated by $< f, g, h >$ and the variable $\mathbf{x}$. With the aid of the factorization relationship, the following can be obtained:

$$\varphi(\mathbf{x}) = < f, g, h >= x_i[fg \dot{+} \overline{f}h](1_i, \mathbf{x}) \dot{+} \overline{x}_i[fg \dot{+} \overline{f}h](0_i, \mathbf{x}).$$

This relationship can also be written as:

$$\varphi(\mathbf{x}) =< f, g, h >= \left\langle x_i, < f_{x_i=1}, g_{x_i=1}, h_{x_i=1} >, < f_{x_i=0}, g_{x_i=0}, h_{x_i=0} > \right\rangle$$

where $f_{x_i=1} = f(1_i, \mathbf{x})$.

This latter relationship enables us to obtain recursively the BDD of a Boolean function without using the graphs. In this case, we choose x_i with the smallest order.

9.2.3 *Probabilistic calculation*

The probabilistic calculation, which will be carried out, will yield the same value in all the cases. However, there is one problem that still remains and it concerns the size of the BDD. Different orders of variables give different sizes. In certain cases, this difference can be very large, which implies an unmatched treatment. In important applications, it will be in our interest to choose correctly the order of the variables. The problem regarding the choosing of the order of variables still remains unsolved.

The calculation for $Q(\mathbf{q}) = E\varphi(\mathbf{X})$ being carried out on the basis of the BDD of Figure 9.3(c), we have:

$$Q(q_1, q_2, q_3) = q_1 q_2 + q_1 p_2 q_3 + p_1 q_3.$$

On the basis of the BDD in Figure 9.4, we will obtain:

$$Q(q_1, q_2, q_3) = q_3 + q_1 q_2 p_3.$$

It is easy to verify, keeping in mind the relationship $q_i + p_i = 1$, that the above two expressions are equal.

9.3 Probabilistic assessment of the FTs based on the BDD

The construction of the BDD for a fault tree does not differ from that of an explicit Boolean function as in the previous section, but for the mode of calculating the value of the Boolean variable for the top event.

We will be considering the fault tree in Figure 9.7 and constructing the corresponding BDD with a view to calculating the probability of the top event and after the BDD for obtaining the prime implicants.

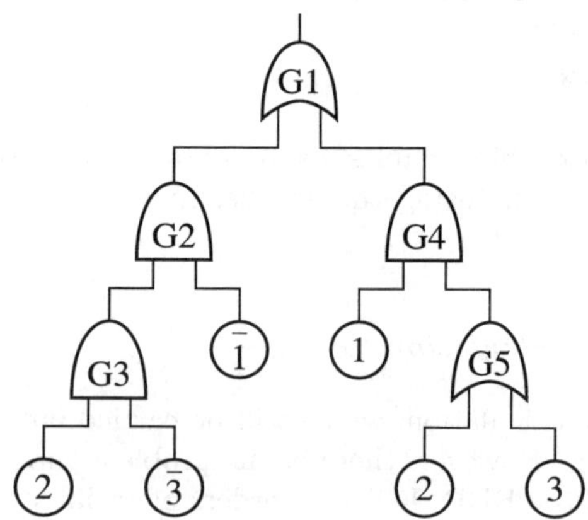

Figure 9.5 *Fault tree*

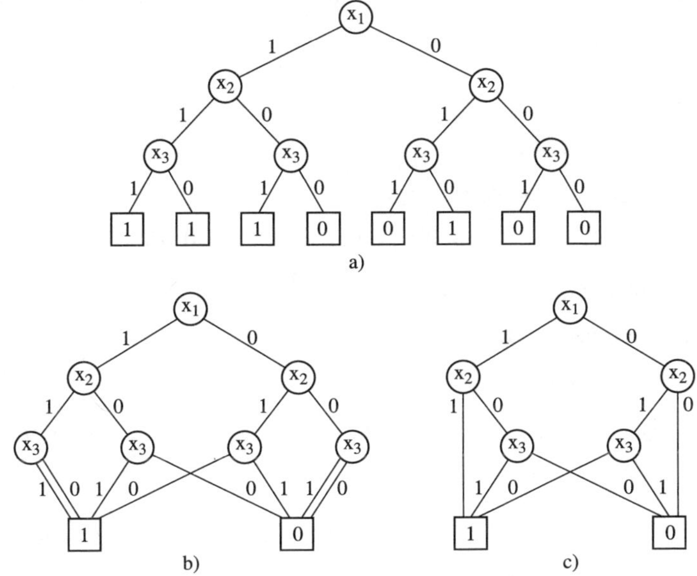

Figure 9.6 *BDD of the fault tree*

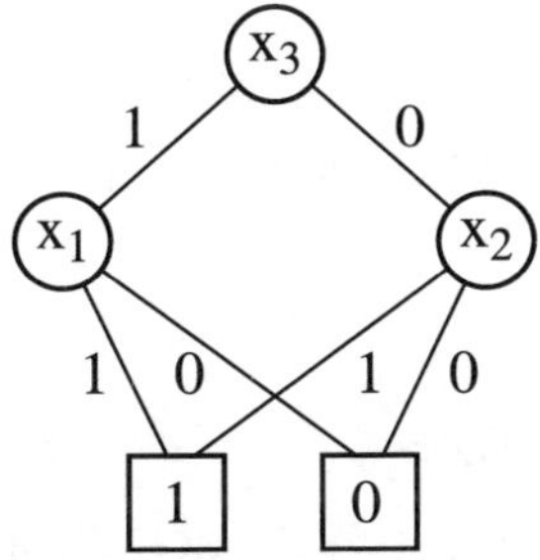

Figure 9.7 *BDD corresponding to a different order of the variables*

(a) Construction of the BDD: it is obtained successively in Figures 9.6(a), (b) and (c).

On adopting a different order: $x_3 < x_1 < x_2$, the BDD that is obtained is given in Figure 9.7 It is seen that it is smaller (3 vertices) than that in Figure 9.6(c) (5 vertices).

(b) Formally obtaining the BDD: from the formal viewpoint, we have:

$$
\begin{aligned}
x_i &= < x_i, 1, 0 >, \\
\overline{x}_i &= < x_i, 0, 1 >, \\
G_3 &= x_2\overline{x}_3 = < x_2, \overline{x}_3, 0 > \\
&= << x_2, 1, 0 >, < x_3, 0, 1 >, 0 > \\
&= < x_2, < 1, < x_3, 0, 1 >, 0 >, < 0, < x_3, 0, 1 >, 0 >> \\
&= < x_2, < x_3, 0, 1 >, 0 >, \\
G_5 &= x_2 \dot{+} x_3 = < x_2, 1, x_3 > \\
&= << x_2, 1, 0 >, 1, < x_3, 1, 0 >> \\
&= < x_2, << 1, 1, 0 >, 1, < x_3, 1, 0 >>, << 0, 1, 0 >, 1, < x_3, 1, 0 >>> \\
&= < x_2, 1, < x_3, 1, 0 >> \\
&= < x_2, 1, x_3 >, \\
G_2 &= \overline{x}_1 G_3 = < x_1, 0, G_3 > \\
&= << x_1, 1, 0 >, 0, G_3 > \\
&= < x_1, < 1, 0, G_3 >, < 0, 0, G_3 >> \\
&= < x_1, 0, G_3 >, \\
G_4 &= x_1 G_5 = < x_1, G_5, 0 > \\
&= < x_1, < 1, G_{5|x_1=1}, 0 >, < 0, G_{5|x_1=0}, 0 >> \\
&= < x_1, G_{5|x_1=1}, 0 > \\
&= < x_1, G_5, 0 >, \\
G_1 &= G_2 \dot{+} G_4 = < G_2, 1, G_4 > \\
&= < x_1, < G_{2|x_1=1}, 1, G_{4|x_1=1} >, < G_{2|x_1=0}, 1, G_{4|x_1=0} >> \\
&= < x_1, < 0, 1, < x_2, 1, x_3 >>, << x_2, < x_3, 0, 1 >, 0 >, 1, 0 >> \\
&= < x_1, < x_2, 1, x_3 >, < x_2, < x_3, 0, 1 >, 0 >> .
\end{aligned}
$$

The BDD, as given in Figure 9.6(c), corresponds well to the expression of G_1, described above, which is finally:

$$x_1x_2 + x_1\overline{x}_2x_3 + \overline{x}_1x_2\overline{x}_3$$

(c) Probabilistic assessment: the calculation for the probability of the top event of the fault tree is done either on the graph of the BDD with the aid of the paths leading to value 1 (vertex 1), or with the help of the expression of G_1 described above.

From the BDD of Figure 9.6(c), we obtain

$$P_S = q_1 q_2 + q_1 p_2 q_3 + p_1 q_2 p_3,$$

and from the BDD of Figure 9.7

$$P_S = q_1 q_3 + q_2 p_3.$$

which comes to the same thing.

9.4 Research about the prime implicants

Let us consider a structure function φ, and let the Shannon development be as follows:

$$\varphi(\mathbf{x}) = x_i \varphi(1_i, \mathbf{x}) \dot{+} \overline{x}_i \varphi(0_i, \mathbf{x}).$$

Taking into account the consensus, this relationship can also be written as follows:

$$\varphi(\mathbf{x}) = x_i \varphi(1_i, \mathbf{x}) \dot{+} \overline{x}_i \varphi(0_i, \mathbf{x}) \dot{+} \varphi(1_i, \mathbf{x}) \varphi(0_i, \mathbf{x}).$$

If φ is monotonic with respect to x_i, then, $\varphi(1_i, \mathbf{x}) \geq \varphi(0_i, \mathbf{x})$, which implies that $\varphi(1_i, \mathbf{x}) \varphi(0_i, \mathbf{x}) = \varphi(0_i, \mathbf{x})$, which in turn enables the representation of φ as follows:

$$\varphi(\mathbf{x}) = x_i \varphi(1_i, \mathbf{x}) \dot{+} \varphi(0_i, \mathbf{x}).$$

If, on the contrary, φ is monotonic with respect to $\overline{x}_i = 1 - x_i$, by the same way, φ is written as:

$$\varphi(\mathbf{x}) = \varphi(1_i, \mathbf{x}) \dot{+} \overline{x}_i \varphi(0_i, \mathbf{x}).$$

Noting by $impl(f)$ the set of prime implicants of the Boolean function φ, we have:

$$impl(\varphi) = P \cup Q \cup R,$$

where the sets P, Q and R are, in the first case:

$$\begin{aligned} P &= impl(\varphi(0_i, \mathbf{x})) \\ Q &= \{\{x_i\} \cup \sigma \mid \sigma \in impl(\varphi(1_i, \mathbf{x}))) \setminus P\} \\ R &= \emptyset. \end{aligned}$$

In the second case we have:

$$P = impl(\varphi(1_i, \mathbf{x}))$$
$$Q = \emptyset$$
$$R = \{\{\overline{x}_i\} \cup \sigma \mid \sigma \in impl(\varphi(0_i, \mathbf{x}))) \setminus P\}.$$

For the general case, where φ is monotonic neither for x_i, nor for $1 - x_i$, we have:

$$P = impl(\varphi(1_i, \mathbf{x})\varphi(0_i, \mathbf{x}))$$
$$Q = \{\{x_i\} \cup \sigma \mid \sigma \in impl(\varphi(1_i, \mathbf{x}))) \setminus P\}$$
$$R = \{\{\overline{x}_i\} \cup \sigma \mid \sigma \in impl(\varphi(0_i, \mathbf{x}))) \setminus P\}.$$

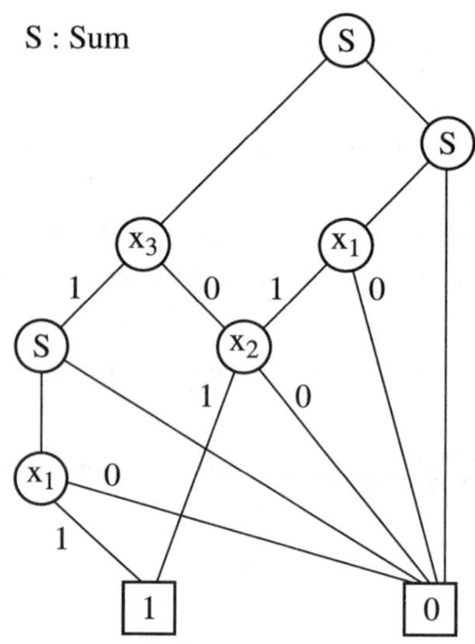

Figure 9.8 *BDD coding the prime implicants of the fault tree*

Concerning the prime implicants, many algorithms have been proposed ([COU 94], [RAU 93], [ODE 95]).

Odeh has proposed an algorithm based on the above analysis [ODE 95], [ODE 96]. It starts from the BDD of the fault tree and from a new type of vertex, called "sum", and constructs a new BDD coding the prime implicants.

The "sum" vertex will realize quite simply the union of the prime implicants of two functions or more precisely of two portions of a BDD (see Appendix A).

This last algorithm, applied to the example of the fault tree in the previous section furnishes the BDD of the prime implicants in Figure 9.8.

The prime implicants are represented by the paths of the BDD leading from the root of the tree at the vertex (1). We obtain the following prime implicants:

$$x_1x_3, \quad x_2\overline{x}_3, \quad x_1x_2.$$

9.5 Algorithmic complexity

The complexity of a BDD is measured by its size, that is, by the number of the vertices contained in it.

The size of a tree developed by Shannon, representing a Boolean function of n variables, is

$$2^0 + 2^1 + 2^2 + ... + 2^n = 2^{n+1} - 1.$$

The maximum size of a reduced Shannon tree or BDD, representing a Boolean function of n variables, is given by:

$$\frac{2^n}{n}(2 + \varepsilon).$$

with $\varepsilon \cong 3.125$. This boundary was obtained by Liaw and Lin [LIA 92] by considering uniquely the reductions due to the equivalent vertices. An improvement on this boundary could be obtained by also considering the reductions of the vertices due to the fact that their two children (1) and (0) head for the same vertex.

Chapter 10

Stochastic Simulation of Fault Trees

10.1 Introduction

The stochastic simulation or the Monte Carlo method is the first technique employed for obtaining, not only the probability of the top event, but also the cut sets of the fault trees. Even though the analytical methods presently developed respond suitably to the calculation requirements of the FTs, it would be of interest to include "stochastic simulation", because simulation offers the advantage of being capable of carrying out the calculations of FTs with less software support. In general, for the problems of the engineer concerning the simulation, the reader can consult for example [BOU 86]. A good description of the FTs is given in [LIE 75]. [HEN 81] and [KUM 77] also treat the simulation of the reliability of the systems modeled by their minimal cut sets.

The theoretical elements necessary for the comprehension of this chapter are included in Appendix C.

10.2 Generation of random variables

In order to be in a position to simulate an FT, we have to be able to generate realizations of given r.v.s.

10.2.1 *Generation of a uniform variable*

The generation of the random variables through methods such as a game of dice, drawing of cards or the stopping of a rotating wheel are already outdated.

The generation of a uniform random variable over $[0,1]$ is presently carried out using a computer in a deterministic procedure; because of this, we call them *pseudo random* variables.

The method that is most often used is based on the congruent sequences. For example, proceeding from an initial value x_0, the foot, we construct the sequence of numbers $x_1, x_2, \ldots$ with the formula

$$x_n = ax_{n-1} + b \mod m, \tag{10.1}$$

where a, b and m are integer constants. For example, we have the values $a = 7^5$, $b = 0$, and $m = 2^{31} - 1$ for a calculator of 32 bits. Thus the values obtained are integers from 0 to $m-1$. Now dividing these values by m, we obtain real values in $[0,1]$. This algorithm for obtaining the random variables is called *a generator of pseudo-random numbers.*

Testing of a generator: we will now present the use of a test of χ^2 for testing whether it is good, i.e., if it really generates the random variables $\mathcal{U}(0,1)$.

We will divide the interval $[0,1]$ into r regular sub-intervals, i.e. $I_i = [\frac{i-1}{r}, \frac{i}{r}[$, $1 \leq i \leq r$.

We will consider a sequence of random variables $U_k, 1 \leq k \leq N$, whose realizations are produced by the generator in question. We define the random variables

$$N_i := \sum_{k=1}^{N} \mathbf{1}_{\{U_k \in I_i\}}, \quad 1 \leq i \leq r. \tag{10.2}$$

Let the r.v. T_N be defined as follows:

$$T_N := \sum_{i=1}^{r} \frac{(N_i - N/r)^2}{N/r}, \tag{10.3}$$

where $1/r$ is the probability that a realization $U_k(\omega)$ falls within the given interval I_i.

We have the following theorem.

Proposition 10.1 *We have the convergence in law*

$$T_N \xrightarrow{\mathcal{L}} \chi^2(r-1), \quad N \to \infty.$$

Thus, for sufficiently large N, we can admit that approximately $T_N \sim \chi^2(r-1)$. For a realization t_N from T_N, and for a level α of risk (in general $\alpha = 0.05$ or 0.01), if $t_N > \chi^2_{1-\alpha}(r-1)$, we reject the generator, and we accept it in the case to the contrary.

In practice, we consider a number of classes verifying: $r \cong N^{2/5}$.

In fact, the method of simulation of the any r.v, which we will be describing in the following discussion, is based on the fundamental theorem (see BOU 86):

Proposition 10.2 *For any r.v X with values in $\mathbb{R}^d$, $d \geq 1$, there exists a mapping g: $\mathbb{R}^n \to \mathbb{R}^d$, and a vector $(U_1, ..., U_n)$ uniformly distributed over the cube $[0,1]^n$, such that the r.v $Y := g(U_1, ..., U_n)$ follows the same distribution as X. In addition, it can be realized by substituting $n = 1$ or $n = d$, or even $n \geq 1$ as given.*

10.2.2 *Generation of discrete random variables*

The problem here consists of generating a d.r.v from the realizations of uniform random variables over $[0,1]$.

Bernoulli random variable: let X be a Bernoulli random variable, $X \sim \mathcal{B}(p)$, $p \in [0,1]$, i.e.

$$\mathbb{P}(X=1) = 1 - \mathbb{P}(X=0) = p.$$

Let there be a random variable $U \sim \mathcal{U}(0,1)$. Then the r.v $Y = \mathbf{1}_{\{U \leq p\}}$ follows a distribution $\mathcal{B}(p)$. In fact,

$$\mathbb{P}(Y=1) = \mathbb{P}(\mathbf{1}_{\{U \leq p\}} = 1) = \mathbb{P}(U \leq p) = p.$$

Any discrete random variables: let there be a d.r.v X with values in $E = \{x_1, x_2, ..., x_n, ...\}$, of distribution $p = (p_1, p_2 ..., p_n, ...)$, i.e., $p_i = \mathbb{P}(X = x_i)$. Let there now be an r.v. $U \sim \mathcal{U}(0,1)$. Then the r.v. Y defined by

$$Y = \sum_{i=1}^{\infty} x_i \mathbf{1}_{\{p_0 + p_1 + \ldots + p_{i-1} < U \leq p_1 + \ldots + p_{i-1} + p_i\}}, \tag{10.4}$$

where $p_0 = 0$, follows the distribution p.

10.2.3 *Generation of real random variables*

We will at first present a general method for the generation of the r.r.v called the *inverse function method*, which is based on the following results.

Proposition 10.3 *Let there be an r.r.v X of c.d.f F continuous. Then the r.v.*

$$Y := F(X) \sim \mathcal{U}(0,1).$$

PROOF. As

$$\{X \leq x\} \subset \{F(X) \leq F(x)\},$$

and

$$\{X > x\} \cap \{F(X) < F(x)\} = \emptyset,$$

We get:

$$\begin{aligned}\mathbb{P}(F(X) \leq F(x)) &= \mathbb{P}(F(X) \leq F(x), X \leq x) + \mathbb{P}(F(X) \leq F(x), X > x)\\ &= \mathbb{P}(X \leq x) + \mathbb{P}(F(X) = F(x), X > x)\\ &= \mathbb{P}(X \leq x).\end{aligned}$$

As F is continuous, for all $y \in]0,1[$, there exists an $x \in \mathbb{R}$ such that $F(x) = y$. Then, from the previous relationship, we have:

$$\mathbb{P}(Y \leq y) = \mathbb{P}(F(X) \leq F(x)) = F(x) = y.$$

Hence the result.

Proposition 10.4 *Let there be an c.d.f F. Let us define $F^{-1} :]0,1[\to \mathbb{R}$, through*

$$F^{-1}(y) := \inf\{x : F(x) \geq y\}, \quad 0 < y < 1.$$

If $U \sim \mathcal{U}(0,1)$, then $X := F^{-1}(U)$ has for r.f. F.

PROOF. Let us show that for any $x \in \mathbb{R}$, such that $F(x) \in]0,1[$, and for any $y \in]0,1[$, $F(x) \geq y$ if and only if $x \geq F^{-1}(y)$.

Let us assume $x \geq F^{-1}(y) = \inf\{x : F(x) \geq y\}, \quad 0 < y < 1$. Then, as F is ascending and continues to the right, $\{x : F(x) \geq y\}$ is an interval containing the extreme left point, and hence $F(x) \geq y$. Now, let us assume $F(x) \geq y$. Then $x \geq \inf\{x : F(x) \geq y\} = F^{-1}(y)$.

As a result:

$$\mathbb{P}(F^{-1}(U) \leq x) = \mathbb{P}(U \leq F(x)) = F(x),$$

which completes the proof.

▷ **Example 10.1.** Let X be an r.v of exponential distribution of parameter $\lambda > 0$, i.e. $X \sim \mathcal{E}$. Let us note F the c.d.f of X and $U \sim \mathcal{U}(0,1)$. The inverse function F^{-1} of F (λ):

$$F^{-1}(y) = -\frac{1}{\lambda}\ln(1-y).$$

Then the r.v. $Y := -\frac{1}{\lambda}\ln(1-U)$, or which arrives at the same relationship, $Y := -\frac{1}{\lambda}\ln(U) \sim \mathcal{E}(\lambda)$

We will also provide another method for obtaining the realizations of absolutely continuous r.v.; we are dealing here with the rejection algorithm of Von Neumann. Let there be a random variable X of p.d.f $f(x)$, $x \in \mathbb{R}$. Let us assume that there exist a p.d.f $g(x)$, $x \in \mathbb{R}$, and a constant $c \geq 1$ such that $cg(x) \geq f(x)$, for all $x \in \mathbb{R}$, for which we will assume that we possess a series of its realizations (Y_n), or that its realizations are easy to obtain [FLU 90].

Algorithm 10.1. Von Neumann rejection method

1. Generate Y from g;
2. Generate U from the uniform distribution over $[0, cg(Y)]$;
3. If $U \leq f(Y)$, then: $X := Y$;
   If otherwise, go back to the stage 1.

End.

In general, g is a uniform distribution.

10.3 Implementation and evaluation of the method

10.3.1 *The Monte Carlo method*

The Monte Carlo method consists of calculating the mathematical expectations for the r.v. functions, i.e.

$$\mathbb{E}\varphi(\mathbf{X}).$$

Let $\mathbf{X}^i, 1 \leq i \leq n$ be the n independent realizations of $\mathbf{X}$. Let us substitute $Z_i = \varphi(\mathbf{X}^i)$, $1 \leq i \leq n$. By the strong distribution of large numbers, we have:

$$\widehat{P}_{S,n} = \frac{1}{n}(Z_1 + ... + Z_n) \xrightarrow{p.s.} \mathbb{E}\varphi(\mathbf{X}), \quad n \to \infty. \tag{10.5}$$

The Monte Carlo method converges:

– at mean value as $O(n^{-1/2})$;

– at worst as $O\Big(\big(\frac{\log\log n}{n}\big)^{1/2}\Big)$,

when $n \to \infty$, [GIR 01].

10.3.2 *Estimating the probability of the top event*

We will estimate the probability of the top event of an FT for a fixed arbitrary time period θ.

Let us consider an FT with $\mathcal{E} = \{e_1, ..., e_N\}$, the set of its basic events, and the r.v. $\mathbf{T} = (T_1, ..., T_N)$ with values in $\mathbb{R}_+^N$, where T_i is the of waiting time for the occurrence of the event i.

Let us note that $f_{\mathbf{T}}$ is the density of $\mathbf{T}$ and φ the structure function of the FT. Let us imagine the case where the system modeled through the FT is not repairable, and let us consider the following application:

$$h : \mathbb{R}_+^N \ni \mathbf{T} \longrightarrow \mathbf{X} = h(\mathbf{T}) \in \{0,1\}^N,$$

with

$$X_i = \mathbf{1}_{\{T_i \le \theta\}}, \quad i = 1, ..., N. \tag{10.6}$$

The estimator of $\widehat{P}_{S,_n}$, the probability of the top event of the FT, P_S for n simulations, and for the time θ, will be given by the following relationship:

$$\widehat{P}_{S,_n} = \frac{1}{n}\sum_{i=1}^{n}\varphi(\mathbf{X}^i), \tag{10.7}$$

where $\mathbf{X}^i$ the ith copy of the r.v. $\mathbf{X} = (X_1, ..., X_N)$. In fact, we are dealing with the empirical estimator, which is unbiased.

In the case of repairable systems, we will proceed as follows:

Let us note $\{T_{ij}, j = 1, 2, ...\}$, the time of appearance with common c.d.f, noted F_i and $\{Y_{i,j}, j = 1, 2, ...\}$, the time of disappearance with common c.d.f, noted G_i, of the basic event $i, i = 1, ..., N$, of the FT.

There exists an $m \in \mathbb{N}^*$ such that:

$$\sum_{k=1}^{m-1} (T_{ik} + Y_{ik}) \leq \theta < \sum_{k=1}^{m-1} (T_{ik} + Y_{ik}) + T_{i,m+1} \tag{10.8}$$

where

$$\sum_{k=1}^{m-1} (T_{ik} + Y_{ik}) + T_{i,m} \leq \theta < \sum_{k=1}^{m} (T_{ik} + Y_{ik}). \tag{10.9}$$

Let $\sum_1^0(\cdot) = 0$.

The realization of the relationship (10.8) is equivalent to the non-occurrence of the basic event i and the relationship (10.9) to the occurrence of the basic event i.

Let us note by L_i the event defined by the relationship (10.9), $\{\omega : \exists m \geq 1$, the relationship (10.9) be realized$\}$. It should also be noted that (10.8) defines the complementary event $\overline{L}_i$ of L_i. Then the relationship (10.6), as described above, is written as follows:

$$X_i = \mathbf{1}_{L_i}, \quad \text{for all} \quad i = 1, ..., N.$$

In this manner, we have, as in the previous case, a realization for $\varphi(\mathbf{X}) = \varphi(X_1, ..., X_N)$. The simulation is terminated, when n realizations of $\varphi(\mathbf{X})$ are obtained.

10.3.3 *Precision of the estimation*

An evaluation per simulation cannot be carried out without an estimation of the error that had occurred.

The precision is described here by two parameters: $\varepsilon > 0$, $0 < \delta < 1$, i.e., if

$$R_n := \widehat{P}_{S,n} - \mathbb{E}\varphi(\mathbf{X}), \tag{10.10}$$

It is expressed as follows:

$$\mathbb{P}(|R_n| \geq \varepsilon) \leq \delta. \tag{10.11}$$

The question that arises is: for a given precision (ε, δ), which is the minimal value of n for expecting this precision?

A first approach that answers this question consists in utilizing the Chebychev inequality. Hence, we can write:

$$P(|\widehat{P}_{S,n} - \mathbb{E}\varphi(\mathbf{X})| \geq \varepsilon) \leq \frac{Var(\varphi(\mathbf{X}))}{n\varepsilon^2}. \tag{10.12}$$

However, for the Bernouilli r.v $\varphi(\mathbf{X})$ we have $Var(\varphi(\mathbf{X})) \leq 1/4$.

As a result, if

$$n \geq n_0 := \left[\frac{1}{4\delta\varepsilon^2} + 1\right] \tag{10.13}$$

the precision expressed by (10.11) will be satisfied, and n_0 defined in (10.13) is the sought-after minimal value. The symbol [a] stands for the integer part of the number a.

A different approach for estimating the minimal value of n is based on the estimation by means of confidence intervals.

Let us substitute: $\gamma := \mathbb{E}[Z]$ and $\sigma^2 := Var(Z)$, $(Z := \varphi(\mathbf{X}))$ and

$$S_n^2 := \frac{1}{n-1}\sum_{i=1}^{n}(Z_i - \widehat{P}_{S,n})^2, \tag{10.14}$$

the estimator of the variance σ^2. By the strong distribution of large numbers, we obtain:

$$S_n^2 \xrightarrow{p.s.} \sigma^2. \tag{10.15}$$

From the central limit theorem, we have:

$$\frac{\sqrt{n}}{\sigma}(\widehat{P}_{S,n} - \gamma) \xrightarrow{\mathcal{L}} \mathcal{N}(0,1). \tag{10.16}$$

From (10.15) and Slutsky's theorem, we obtain

$$\frac{\sqrt{n}}{S_n}(\widehat{P}_{S,n} - \gamma) \xrightarrow{\mathcal{L}} \mathcal{N}(0,1). \tag{10.17}$$

n being sufficiently large, we can consider that approximately

$$\frac{\sqrt{n}}{S_n}(\widehat{P}_{S,n} - \gamma) \sim \mathcal{N}(0,1),$$

and hence

$$\mathbb{P}\Big(z_{\alpha/2} < \frac{\sqrt{n}}{S_n}(\widehat{P}_{S,n} - \gamma) < z_{1-\alpha/2}\Big) = 1 - \alpha$$

or

$$\mathbb{P}\Big(\widehat{P}_{S,n} - z_{1-\alpha/2}\frac{S_n}{\sqrt{n}} < \gamma < \widehat{P}_{S,n} + z_{1-\alpha/2}\frac{S_n}{\sqrt{n}}\Big) = 1 - \alpha,$$

where z_α is the α-fractile of $\mathcal{N}(0,1)$, and we have: $z_\alpha = -z_{1-\alpha}$.

The width of the confidence interval described above is: $l := 2z_{1-\alpha/2}\frac{S_n}{\sqrt{n}}$, and it represents the precision of the approximation.

When the precision is defined by α and δ, we should have $l \leq \delta$, and hence n should verify the following inequality:

$$n \geq (2z_{1-\alpha/2}\frac{S_n}{l})^2. \tag{10.18}$$

$\triangleright$ **Example 10.2.** *Estimation of the mean time until the occurrence of the top event.* Let us consider an FT, whose minimal cut sets are:

$$K_1 = \{1,2,3\} \quad \text{and} \quad K_2 = \{1,4,5\}$$

Let T_i be the time of occurrence of the basic event i, and T the time of occurrence of the top event. We have:

$$T = \min\{\max\{T_1,T_2,T_3\},\max\{T_1,T_4,T_5\}\}.$$

Here, we have the estimator $\widehat{T}_n$ of $\mathbb{E}T$, i.e.

$$\widehat{T}_n := \frac{1}{n}(T^1 + ... + T^n),$$

where T^i, $1 \leq i \leq n$ are the i.i.d. $\sim T$.

Thus, we can write:

$$T_1 \leq T \leq T_1 + T_2 + T_3$$

and hence

$$Var(T) = E[T^2] - (E[T])^2 \leq E[(T_1 + T_2 + T_3)^2] - (E[T_1])^2. \tag{10.19}$$

If $E[T_i] = 1/\lambda_i$, $i = 1,2,3,4$; the relationship (10.19) gives:

$$Var(T) \leq \frac{1}{\lambda_1^2} + \frac{2}{\lambda_2^2} + \frac{2}{\lambda_3^2} + \frac{2}{\lambda_1\lambda_2} + \frac{2}{\lambda_2\lambda_3} + \frac{2}{\lambda_3\lambda_1}.$$

This boundary of variance enables us to calculate n_0 from

$$n_0 = \left[\frac{h}{\delta\varepsilon^2} + 1\right].$$

10.3.4 *Acceleration of the convergence*

We will estimate $\mathbb{E}h(Z)$ based on the Monte Carlo method. The smaller the variance of Z, the faster the convergence will be. This assertion is intuitively true. Instead of considering r.v. Z, we consider another r.v. V such as $Var(V) < Var(Z)$.

Let f be the p.d.f of Z and g that of V. We will assume that $supp(f) \subset supp(g)$ where $supp(.)$ stands for the support of the corresponding functions. Then we can write:

$$\mathbb{E}h(Z) = \mathbb{E}[h(V)L(V)] =: \gamma, \tag{10.20}$$

with

$$L(V) = \frac{f(V)}{g(V)}.$$

On the basis of this equality, we will propose the following estimator for γ:

$$\gamma_n := \frac{1}{n}\sum_{i=1}^{n} h(V_i)L(V_i). \tag{10.21}$$

The variance of this estimator is:

$$\begin{aligned} Var(\gamma_n) &= \frac{1}{n} Var[h(V)L(V)] \\ &= \frac{1}{n}\Big\{\mathbb{E}[h(V)L(V)]^2 - [\mathbb{E}(h(V)L(V)]^2\Big\} \\ &= \frac{1}{n}\Big\{\mathbb{E}[h(V)L(V)]^2 - \gamma^2\Big\}. \end{aligned}$$

The problem that is now posed is: what is the p.d.f (or the r.v V) that minimizes the variance of γ_n?

It is clear that for minimizing the $Var(\gamma_n)$, $\mathbb{E}[h(V)L(V)]^2$ should be minimized. This arrives at the selection for p.d.f

$$g(x) = \frac{1}{\gamma} h(x)f(x). \tag{10.22}$$

In this case, we have:

$$Var(\gamma_n) = 0,$$

and therefore, a single realization would suffice for obtaining the exact value of γ. Evidently, we will find out quickly that this poses a problem, that is, δ is the parameter that we have to estimate.

Nevertheless, in many cases, we are able to define a p.d.f g for reducing the variance and hence accelerate the speed of convergence.

10.3.5 *Rare events*

In most cases, the events involved in an FT have very low probabilities (aircraft crash, explosion of a nuclear power plant, general breakdown of the IT system of a bank, etc., where the probability of such an event is generally of the order of 10^{-10}), and, consequently, the probability of observing an event within a rather short time period is likewise considerably very small. In fact, the simulation described earlier will be very expensive.

We will discuss a problem which is posed during a calculation of the probability of an event with low probability of occurrence.

Let there be such an event expressed by $\{X \in A\}$ and let us set $h = \mathbf{1}_A$. We have:

$$\gamma = \mathbb{E}h(X) = \mathbb{P}(X \in A).$$

Here, we will differentiate between two families of problems: one with bound relative error and the other with unbound relative error.

Relative error: the relative error $RE(X)$ of a random variable X is defined by the following ratio:

$$RE(X) := \frac{\text{standard deviation}}{\text{expectation}} \tag{10.23}$$

For a random variable $X \sim \mathcal{B}(\gamma)$, we have $RE(X) = 1/\sqrt{\gamma}$. For the case of an estimator γ_n of an indicator random variable, shown above, we have $\mathbb{E}\gamma_n = \gamma$ and $Var(\gamma_n) = \gamma(1-\gamma)/n$. Hence

$$RE(\gamma_n) \cong \frac{1}{\sqrt{n\gamma}} \longrightarrow \infty, \quad \gamma \to 0.$$

In this case the relative error is not bounded. These problems are very difficult to handle.

Exercises

Chapter 1

E 1.1. The distribution governing the life duration T of a system is without atom, having support $\mathbb{R}_+$. Show that the probability, whereby a failure occurs at a time which is a fully positive natural number or rational value, is zero.

E 1.2. The distribution Q of the life duration T of a system has a distribution function F, given by the relationship:

$$F(x) = \begin{cases} 1 - \exp\{-\lambda x\} & \text{if } x \geq a, \\ 0 & \text{if } x < a, \end{cases}$$

where $a > 0$, $\lambda > 0$.

a) Calculate $Q(a)$.

b) Calculate the $MTTF$ and the variance of T.

c) Study the characteristics of the magnitudes that are calculated, when $a \to 0+$.

E 1.3. Let R(t) be the reliability function of a component/system, and suppose that there exist a positive number a, such that $\lim_{t\to\infty} e^{at}R(t) = 0$. Show that the MTTF exists and

$$\int_0^\infty R(t)dt.$$

.

E 1.4. The lifetime of a system is the r.v T with c.d.f F on the half-real axis $x \geq 0$. Its mission duration is a v.a. θ of c.d.f G again over $x \geq 0$. Calculate the probability whereby the system will successfully accomplish its mission.

E 1.5. Let there be a component under continuous working. When it breaks down, it is replaced instantaneously by a new component. If the p.d.f for the survival time of the components is $f(t) = \lambda \exp\{-\lambda t\}, t > 0$, give the probability density for the time of the nth failure.

E 1.6. A component taken from a stock containing identical components is put to work. The stock contains the components in good condition as well as the faulty components. When we take out a component without any special precaution, its probability of being a failure is given equal to q. It has been shown that the c.d.f of the lifetime of the working component T is given by the relationship $F(x) = 1 - (1-q)\exp\{-\lambda x\}$, for $x > 0$. Show that the average duration of good functioning of the component is $E[T] = (1-q)/\lambda$.

Chapter 2

E 2.1. Let there be a binary, coherent system $S = (C, \varphi)$, of the order $n > 1$. Show, with respect to the active redundancy, that a local redundancy (redundancy for each component) is more reliable than a global redundancy (redundancy of the system). Find the equivalence condition.

E 2.2. With respect to the passive redundancy, show that: (a) in a system of 1-out-of-n:G, the global redundancy is less reliable than the local redundancy; (b) in a system n-over-n:G, the global redundancy is more reliable than the local redundancy ([SHE 91]).

E 2.3. Show that the reliability function of a binary system of order n, noted as $r(\mathbf{p})$, where $\mathbf{p} = (p_1, ..., p_n)$, is a multiaffine function determined in a unique manner by the 2^n coefficients in the following form:

$$r(\mathbf{p}) = c_0 + \sum_{i=1}^{n} c_i p_i + \sum_{1 \leq i < j \leq n} c_{ij} p_i p_j + ... + c_{12...n} p_1 p_2 \cdots p_n.$$

Show also that this relationship is correct, if, for any path vector, it gives the value 1, and for any cut set vector, it gives the value 0 ([RUS 83]).

E 2.4. Let us consider a parallel system of the order 3. When the three components are working, the failure rate of each of them is given as $\lambda/3$. When one of the three has broken down, the failure rate of the other two is seen to be $\lambda/2$ and, when only one of the three is working, its failure rate is λ. Give the c.d.f of the system's lifetime. The same criterion is considered when the system is in passive redundancy.

E 2.5. Let us consider a system of order 2 and let $f(x, y)$ be the joint probability density of the lifetimes of the two components.

(a) Show that the reliability is given by the following relationships depending on the case:

– series system: $R(t) = \int_t^\infty dx \int_t^\infty f(x, y)dy$;

– active redundancy: $R(t) = 1 - \int_0^t dx \int_0^t f(x, y)dy$;

– passive redundancy: $R(t) = 1 - \int_0^t dx \int_0^{t-x} f(x, y)dy$.

(b) If the lifetimes of the two components are independent, derive the well known expressions for the reliability in the above three cases starting from these three expressions.

(c) Write down the above relationships in the case of a system of order $n > 2$.

Chapter 3

E 3.1. Dealing with the hydraulic system (Figure 3.7), construct the FT for the top event "output less than or equal to 50%".

E 3.2. Propose an algorithm for the automatic construction of an FT starting from block diagram of reliability.

Chapter 4

E 4.1. Consider the fault tree given below, as described by its operators and by their inputs in each line of the following table. For example, $G + 15$ refers to the OR operator number 15, and $G * 6$ refers to the AND operator number 6.

G+1	G+2	02	G+15
G+2	01	G+3	
G+3	G+4	G+5	
G+4	03	04	05
G+5	G*6	G+7	
G*6	G+8	G+12	
G+7	06	07	
G+8	G+9	G+11	
G+9	11	G+10	
G+10	12	13	
G+11	14	15	16
G+12	G+17	G+19	G+13
G+13	G+14	G+10	
G+14	17	18	
G+15	08	G+16	
G+16	09	10	
G+17	22	23	24

Design this FT and give the lists of the minimal cut sets and minimal paths.

E 4.2. Let the following FT be described by its operators and by its inputs:

G+1	01	G*2	06	07
G*2	G+3	G+4		
G+3	02	03		
G+4	04	05		

Design this FT and give the lists of the minimal cut sets and of the minimal paths.

E 4.3. Let there be a coherent FT, whose basic events are $\mathcal{E} = \{e_1, ..., e_N\}$ and let the application $v : e_i \mapsto v(i)$, where $v(i)$ is the i-th prime number. On the basis of the application v, propose a reduction algorithm of the minimal cut sets.

Chapter 5

E 5.1. For the FT of Exercise 4.1:

(a) Calculate the probability of occurrence of the top event, if the probabilities for the occurrence of the basic events are equal to 10^{-4}.

(b) Calculate the four types of bounds.

(c) Use the Hughes method for calculating the error committed concerning the probability of the top event, when its evaluation is done solely from the cut sets of length 1.

E 5.2. For the FT of Exercise 4.2:

(a) Calculate the probability of occurrence of the top event, if the probabilities for the occurrence of the basic events are: $p_1 = 10^{-3}$, $p_2 = 3 \cdot 10^{-2}$, $p_3 = 10^{-2}$, $p_4 = 3 \cdot 10^{-3}$, $p_5 = 10^{-2}$, $p_6 = 3 \cdot 10^{-3}$ and $p_7 = 10^{-6}$.

(b) Calculate the four types of bounds.

(c) Use the Schneeweiss method for calculating the error that occurred for the probability of the top event, when its evaluation is solely done from the cut sets of length 1.

E 5.3. Let there be an FT whose set of repeated events is $R = \{e_{i_1}, ..., e_{i_r}\}$ and whose structure function is φ.

(a) Show that the probability of the top event is given the following formula:

$$P_S = \sum_{(v_{i_1},...,v_{i_r}) \in \{0,1\}^r} E\varphi(v_{i_1}, ..., v_{i_r}, \mathbf{X}) \prod_{k=1}^{r} EX_{i_k}^{\alpha_{i_k}},$$

where

$$X_{i_k}^{\alpha_{i_k}} = \begin{cases} X_{i_k} & \text{if } \alpha_{i_k} = 1 \\ 1 - X_{i_k} & \text{if } \alpha_{i_k} = 0. \end{cases}$$

(b) Propose an algorithm for calculating the P_S on the basis of the relationship (a) [ODE 93].

E 5.4. The appearance (arrival) rates of the two inputs of an AND operator are λ_1 and λ_2.

Show that:

1. The appearance of the output Λ is given by:

$$\Lambda(t).\frac{\lambda_1 e^{-\lambda_1 t}+\lambda_2 e^{-\lambda_2 t}-(\lambda_1+\lambda_2)e^{-(\lambda_1+\lambda_2)t}}{e^{-\lambda_1 t}+e^{-\lambda_2 t}-e^{-(\lambda_1+\lambda_2)t}}.$$

2. If $\lambda_1 \neq \lambda_2$, then

$$\lim_{t\to\infty} \Lambda(t)=\min\{\lambda_1,\lambda_2\}.$$

3. There is a $t_0 \geq 0$ such that, for all $t > t_0$, we have

$$\min\{\lambda_1,\lambda_2\} \leq \Lambda(t) \leq \max\{\lambda_1,\lambda_2\}.$$

4. If $\lambda_1 = \lambda_2 = \lambda$ then

$$\lim_{t\to\infty} \Lambda(t)=\lambda.$$

5. Generalize the formula of 1) for n inputs.

Chapter 6

E 6.1. For the FT of Exercise 5.1, calculate the Birnbaum's factor of importance for the events 01 and 11.

E 6.2. Let us consider a coherent, binary system $S=(C,\varphi)$ of the order $n \geq 1$. Let us note $\mathbf{p}=(p_1,\dots,p_n)$ as being the reliabilities of its components. If we replace any component i of the system by two components identical to the component i in parallel, show that:

(a) The growth of the reliability of the system will be $\Delta R^i = p_i q_i I_B^{(i)}$, where $q_i = 1-p_i$ and $I_B^{(i)}$ are the Birnbaum's probabilistic importance factor.

(b) The growth of the reliability satisfies the relationship:

$$\frac{c_i}{1+c_i}R(\mathbf{p})q_i \leq \Delta R^i \leq \min\{R(\mathbf{p})q_i, p_i(1-R(\mathbf{p}))\},$$

where $c_i = p_i I_B^{(i)}$ ([SHE 90]).

E 6.3. Let us consider a coherent, binary system $S=(C,\varphi)$ of the order $n \geq 1$.

(a) Show that the probability for the component i having caused the failure of the system at time t, is given by:

$$\frac{[R(1_i, \mathbf{p}(t)) - R(0_i, \mathbf{p}(t))] f_i(t)}{\sum_{j=1}^{n} [R(1_j, \mathbf{p}(t)) - R(0_j, \mathbf{p}(t))] f_j(t)}.$$

(b) Deduce from the above relation the importance factor according to Barlow-Proschan ([BAR 75]).

E 6.4. Assume that P_S follows a Beta distribution and give a method for the assessment of the uncertainty proceed in a manner analogous to the method of empirical distribution; see section 6.1.2. The density of the Beta distribution with parameters a and b ($a > 0, b > 0$) is

$$f(x) = \frac{1}{B(a,b)} x^{a-1}(1-x)^{b-1}, \quad 0 < x < 1, a > 0, b > 0,$$

where $B(a,b)$ is the Beta function defined by

$B(a,b) = \int_0^1 x^{a-1}(1-x)^{b-1} dx$, and we also have $B(a,b) = \Gamma(a)\Gamma(b)/\Gamma(a+b)$.

Chapter 7

E 7.1. Let us consider a binary system of order 3 that has to function in three consecutive phases of respective durations: θ_1, θ_2 and θ_3.

The structures of the system corresponding to the three phases are:

Phase 1: 1//2//3 (parallel system)

Phase 2: 1 – (2//3)

Phase 3: 1 – 2 – 3 (series system)

(a) Construct the FT concerning the failure of the system's mission.

(b) Find the minimal cut sets of the FT.

(c) Calculate the probability of occurrence of the top event.

(d) Numerical application: $\theta_1 = \theta_3 = 100$ h, $\theta_2 = 300$ h, $\lambda_1 = \lambda_2 = \lambda_3 = 10^{-3}$ h^{-1}.

E 7.2. Carry out a modular decomposition of the FT in Exercise 4.1 through the Olmos-Wolf algorithm. Next, calculate the boundaries on the basis of the preceding decomposition and compare the results with those found in Exercise 5.1.

Chapter 8

E 8.1. Let us consider the system: $S = (C, [0,1]^n, [0,1], \varphi)$. The (s, r)-importance of the component i for the level z is defined by the relationship:

$$I_i^{(s,r)}(z) = P\{\varphi(s_i, \mathbf{X}) \geq z\} - P\{\varphi(r_i, \mathbf{X}) \geq z\},$$

where $s, r, z \in [0,1]$.

Show that the $(0,1)$-importance is given by:

$$I_i = E[\varphi(1_i, \mathbf{X})] - E[\varphi(0_i, \mathbf{X})]$$

and

$$\Delta E\varphi(\mathbf{X}) = I_i \Delta E[X_i].$$

E 8.2. Let us consider a discrete multi-performance system: $S = (C, \mathcal{E}^n, \mathcal{E}, \varphi)$, show that:

(i) $\varphi(\mathbf{x} \dot{+} \mathbf{y}) \geq \varphi(\mathbf{x}) \dot{+} \varphi(\mathbf{y})$,

(ii) $\varphi(\mathbf{xy}) \leq \varphi(\mathbf{x})\varphi(\mathbf{y})$ where $\mathbf{x}, \mathbf{y} \in \mathcal{E}^n$ ([BLO 82]).

Chapter 9

E 9.1. Let there be an FT as given below.

G+1	G*2	G*3	G*4
G*2	1	2	
G*3	$\overline{1}$	3	
G*4	$\overline{3}$	4	

(a) Construct the BDD of this tree.

(b) Show that the prime implicant are: $\{2, 4\}$, $\{2, 3\}$, $\{\overline{3}, 4\}$, $\{1, 2\}$, $\{\overline{1}, 4\}$ and $\{\overline{1}, 3\}$.

(c) If $p_1 = 0,1$; $p_2 = 0,2$; $p_3 = 0,3$ et $p_4 = 0,4$, show that the probability of the top event is equal to $0,573$ [ODE 95].

E 9.2. Construct the BDD of the FT of Exercise 4.1.

E 9.3. Construct the BDD of the FT of Exercise 4.2.

Chapter 10

E 10.1. Starting from the realizations of the r.v with c.d.f f.r. F_i, $i = 1, ..., n$, generate the realizations of the random variable with c.d.f F:

1. $F(x) \equiv \prod_{k=1}^{n} F_k(x)$ (series system).
2. $F(x) \equiv 1 - \prod_{k=1}^{n}(1 - F_k(x))$ (parallel system).

E 10.2. Show that the exponential distribution has a bound relative error.

E 10.3. Write an algorithm for obtaining the minimal cut sets of a coherent fault tree (start from the algorithm in section 9.4; see algorithm A5).

APPENDICES

Appendix A

BDD Algorithms in FT Analysis

A1 Introduction

In order to generate a BDD corresponding to the structure function of an FT, we study the FT in depth, and we recursively construct the BDD for each node.

Section A2 presents the construction of a BDD starting from an FT. Algorithm 1, for this construction, takes recourse to algorithm 2. Section A3 presents the probabilistic assessment (algorithm 3) of the top event (or of an intermediate event), proceeding from the BDD constructed beforehand. Section A4 presents the calculation of the Birnbaum importance factor for the top event (algorithm 4) or of an intermediate event. Section A5 presents algorithm 5 for searching the prime implicants for a non-coherent FT, or the minimal sets (cut sets and paths) for a coherent FT. Algorithm 5 in combination with algorithm 6 completes the difference between two Boolean functions (BDD) F and G, i.e., $F \setminus G$. It generates a BDD coding for all the paths of F (as indicated in the graph), except those containing a path of G. Finally, algorithm 7 reads the reduced BDD and displays the prime implicants.

These algorithms (1–7) can constitute the basis of a software for the analysis of FTs.[1]

1 Written by Khaled Odeh.

A2 Obtaining the BDD

In this section, we present an algorithm for the construction of the BDD from the FT. The FT-to-BDD (node) function transforms the BDD of the FT, whose top event is "node".

Algorithm 1. *Construction of the BDD from the FT*

```
FT-to-BDD(node)
Start
  If (node is a basic event) then
    R := ite(node, 1, 0)
  if not /* node is an operator */
    op := the operator associated with node
    j := first children of node
    R := FT-to-BDD(j)
    For (for all the threads i of node and i ≠ j) Make
      F := FT-to-BDD(i)
      R := BDD-OP(R, F, op)
    End for
  End if
  Return(R)
End
```

The function "BDD-OP (F, G, op)" constructs the BDD associated with the Boolean operation "op" between F and G. This function transforms the triplet "(op, F, G)" from the following table into $ite(F,G,H) \equiv < F,G,H >$, ("ite" is an abbreviation of "if-then-else").

Operator	ite
$F \cdot G$	ite<F, G, 0>
$F \dot{+} G$	ite<F, 1, 0>
$F \oplus G$	ite<F, $\overline{G}$, 0>

The latter will then be assessed by applying algorithm 2. Let us note that algorithm 2 calls on the functions linked with the hashing table. We will use two tables:

1. A table of operations where the Boolean operations between two BDDs are memorized. This helps to avoid repeated calculations that have already been carried out.

2. A table in which the diverse nodes of the BDD are memorized. This helps to avoid the creation of already existing "ite"s.

These different functions are as follows:

1. "exist-op-table (<F,G,H>, R)" returns R in the case where there exists an input at <F,G,H> in the table of operations.

2. "add-op-table (<F,G,H>, R)" creates an input $< F, G, H >$ in the table of operations for storing the BDD R.

3. "add-or-restore-ite-table ($< x, F_{x=1}, F_{x=0} >, F$)" resets F, if there is an input at $< x, F_{x=1}, F_{x=0} >$ in the table of the "ite"s; otherwise it creates for itself an input for storing F.

Algorithm 2. *Algorithm of binary operation between two BDDs*

```
BDD-OP(F, G, op)
Start
 If (F = 1 or F = 0) and (G = 1 or G = 0) then
   Return(F op G)
 If not
   R := T[F, G]
   If (R ≠ NULL) /* the BDD of F op G is already calculated*/
     Return(R)
   If not
     R :=create a new BDD
     T[F, G] := R /* store the result in the table T */
     index(R) := min(index(F), index(G))
     If (index(F) = index(R)) then
       L1 := F_{r(F)=1}; H1 := F_{r(F)=0}
     Otherwise
       L1 := F; H1 := F
     End if
     If (index(G) = index(R)) then
       L2 := G_{r(G)=1}; H2 := G_{r(G)=0}
     Otherwise
       L2 := G; H2 := G
     End if
     R_{r(R)=1} :=BDD-OP-1(L1, L2, op)
     R_{r(R)=0} :=BDD-OP-1(H1, H2, op)
     Return(R)
   End if
 End if
End
```

A3 Algorithm of probabilistic assessment

We will present an algorithm of assessment of the probability of top event from the BDD of the FT.

Algorithm 3. *Probabilistic assessment of the BDD*

```
Probability(F)
Start
  If (F = 0) then
    Return(0)
  Otherwise
    If (F = 1) then
      Return(1)
    Otherwise
      If (exists-prob-table(F, R))
        Return(R)
      Otherwise
        R := p_r(F) · Probability(F_r(F)=1) + q_r(F) · Probability(F_r(F)=0)
        add-prob-table(F, R)
        Return(R)
      End if
    End if
  End if
End
```

A4 Importance factors

This algorithm calculates the importance factor of the intermediate event F. When $F = Top$, then it is the top event.

Algorithm 4. *Calculation of the Birnbaum importance factor*

```
Importance(F, i)
Start
  If(F = 1 or F = 0) then
    Return(0)
  Otherwise
    If (exists-importance-table(F, R))
      Return(R)
    Otherwise
      If (index(F) = index(x_i))
      /* We will recuperate the probabilities of table-prob */
        exists-prob-table(F_{x_i=1}, R1)
        exists-prob-table(F_{x_i=0}, R2)
        R := R1 − R2
      Otherwise
        If (index(F) < index(x_i))
          R = P_{r(F)}·Importance(F_{r(F)=1}, i)
                      +
              q_{r(F)}·Importance(F_{r(F)=0}, i)
        Otherwise
          If (index(F) > index(x_i))
              Return(0)
        End if
      End if
      add-importance-table(F, R)
      Return(R)
    End if
  End if
End
```

A5 Prime implicants

Algorithm 5. *Obtaining the BDD of the prime implicants of F*

Imp(F)
Start
If $(F=0$ or $F=1)$ $then$ $Return(F)$
$Otherwise$
If $(exists\text{-}implic\text{-}prem\text{-}table(F,\widetilde{F}))$ $Return(\widetilde{F})$
$Otherwise$
$non\_monotone := 0$
$\widetilde{F}_{r(F)=0} := Imp(F_{r(F)=0})$
$\widetilde{F}_{r(F)=1} := Imp(F_{r(F)=1})$
If (f is monotonic with respect to $r(F)$) $then$
$\widetilde{F}_{r(F)=1} := without(\widetilde{F}_{r(F)=1}, \widetilde{F}_{r(F)=0})$
$Otherwise$
If (f is monotonic with respect to $\overline{r(F)}$) $then$
$\widetilde{F}_{r(F)=0} := sans(\widetilde{F}_{r(F)=0}, \widetilde{F}_{r(F)=1})$
$Otherwise$
$non\_monotone := 1$
/* we will calculate $F_{r(F)=1} \wedge F_{r(F)=0}$ */
$cons := ite(F_{r(F)=1}, F_{r(F)=0}, 0)$
$cons := Imp(cons)$ /* we will calculate the consensus*/
$\widetilde{F}_{r(F)=1} := without(\widetilde{F}_{r(F)=1}, cons)$
$\widetilde{F}_{r(F)=0} := without(\widetilde{F}_{r(F)=0}, cons)$
End if
End if
$add\text{-}or\text{-}recover\text{-}ite\text{-}table(< r(F), \widetilde{F}_{r(F)=1}, \widetilde{F}_{r(F)=1} >, \widetilde{F})$
If $(non\_monotone = 1)$ $then$
$\widetilde{F} := Sum(\widetilde{F}, cons)$ /* for representing the union of $\widetilde{F}$ and $cons$ */
$add\text{-}prim\text{-}implic\text{-}table(F, \widetilde{F})$
$Return(\widetilde{F})$
End if
End if
End

Algorithm 6. *The BDD of* $F \backslash G$

```
Without(F, G)
Start
  If (F = 0 or G = 1 then F = G) then   Return(0)
  Otherwise
    If (G = 0) then   Return(F)
    Otherwise
      If (F = 1) then   Return(1)
      Otherwise
        If(exists-without-table(F, G, R)) then   Return(R)
        Otherwise
          If (index(F) < index(G)) then
            R_{r(F)=1} := without(F_{r(x)=1}, G)
            R_{r(F)=0} := without(F_{r(x)=0}, G)
          Otherwise
            If (index(F) > index(G)) then
              If (g is monotone with respect to r(g)) then
                Return(without(F, G_{r(G)=0})
              Otherwise
                If (g is monotone with respect to ¬r(g)) then
                  Return(without(F, G_{r(G)=1})
                Otherwise
                  Return(F)
            Otherwise
              If (index(F) = index(G)) then
                If (f is monotone with respect to r(f)) then
                  R_{r(F)=1} := without(without(F_{r(F)=1}), G_{r(F)=1}), G_{r(F)=0})
                  R_{r(F)=0} := without(F_{r(F)=0}, G_{r(F)=0})
                Otherwise
                  If (f is monotone with respect to ¬r(f)) then
                    R_{r(F)=1} := without(F_{r(F)=1}), G_{r(F)=1})
                    R_{r(F)=0} := without(without(F_{r(F)=0}), G_{r(F)=0}), G_{r(F)=1})
                  Otherwise
                    R_{r(F)=1} := without(F_{r(F)=1}), G_{r(F)=1})
                    R_{r(F)=0} := without(F_{r(F)=0}, G)
                  End if
                End if
              End if
            End if
          End if
          add-or-recover-ite-table(< r(F), R_{r(F)=1}, R_{r(F)=1} >, R)
          add-without-table(F, G, R)
        End if
End
```

Algorithm 7. *Display the prime implicants coded by F*

Displ-imp($\widetilde{F}, \sigma$)
Start
  If ($\widetilde{F} = 0$) *then* *Return*
  Otherwise
    If ($\widetilde{F} = 1$) *then* *display*(σ) /* we display the prime implicant σ */
    Otherwise
      If ($\widetilde{F} = Sum(G, H)$) *then*
        /* the prime implicants of $\widetilde{F}$ are those of G union H */
        Displ-imp(G, σ)
        Displ-imp(H, σ)
      Otherwise
        If (f is monotone with respect to $r(\widetilde{F})$) *then*
          /* the prime implicants of $\widetilde{F}$ are those of $\widetilde{F}_{r(\widetilde{F})=1}$
          to which we add $r(\widetilde{F})$ union those of $\widetilde{F}_{r(\widetilde{F})=0}$ */
          Displ-imp($\widetilde{F}_{r(\widetilde{F})=1}, \sigma \cup r(\widetilde{F})$)
          Displ-imp($\widetilde{F}_{r(\widetilde{F})=0}, \sigma$)
        Otherwise
          If (f is monotone with respect to $\overline{r(\widetilde{F})}$) *then*
            /* the prime implicants of $\widetilde{F}$ are those of $\widetilde{F}_{r(\widetilde{F})=0}$
            to which we add $\overline{r(\widetilde{F})}$ union those of $\widetilde{F}_{r(\widetilde{F})=1}$ */
            Displ-imp($\widetilde{F}_{r(\widetilde{F})=0}, \sigma \cup \overline{r(\widetilde{F})}$)
            Displ-imp($\widetilde{F}_{r(\widetilde{F})=1}, \sigma$)
          otherwise
            /* the prime implicants of $\widetilde{F}$ are those of $\widetilde{F}_{r(\widetilde{F})=1}$ to which we
            add $r(\widetilde{F})$ union those of $\widetilde{F}_{r(\widetilde{F})=1}$ to which we add $\overline{r(\widetilde{F})}$ */
            Displ-imp($\widetilde{F}_{r(\widetilde{F})=1}, \sigma \cup r(\widetilde{F})$)
            Displ-imp($\widetilde{F}_{r(\widetilde{F})=0}, \sigma \cup \overline{r(\widetilde{F})}$)
          End if
        End if
      End if
    End if
  End if
End

Appendix B

European Benchmark Fault Trees

B1 Description of the data

Each line of the structural data corresponds to the description of an operator of the fault tree. For example, the line "G*143 G+139 G+140 G+141 G+142" will describe the inputs of the operator G*143, which are: G+139 G+140 G+141 G+142.

We identify the type of operator by the second character:

1. *: for an AND operator, i.e. "G*143".
2. +: for an OR operator, i.e., "G+139".
3. n/k: for a combination operator n/k, i.e., "G(3/4)112".

An event is designated by the letter "T" followed by:

1. N: normal.
2. C: complementary,

and its number, for example "TN252".[1]

---

1 Written by Khaled Odeh.

B2 Fault tree: Europe-1

B2.1 *Structure of the fault tree (structural data)*

G*1	G+126	G+138	G+144		
G+144	G+111	G+112	G*143	TN253	
G*143	G+139	G+140	G+141	G+142	
G+142	G*65	G*69	G+118	G+132	TN251
G+141	G*64	G*67	G+117	G+131	TN250
G+140	G*63	G*68	G+118	G+130	TN249
G+139	G*62	G*66	G+117	G+129	TN248
G+138	G+106	G+119	G+137		
G(3/4)137	G+133	G+134	G+135	G+136	
G+136	G*65	G+132	TN261		
G+135	G*64	G+131	TN260		
G+134	G*63	G+130	TN259		
G+133	G*62	G+129	TN258		
G+132	G*65	G*128	TN257		
G+131	G*64	G*127	TN256		
G+130	G*63	G*128	TN255		
G+129	G*62	G*127	TN254		
G*128	G*116	G+120			
G*127	G*115	G+120			
G+126	G+111	G+112	G+125	TN253	
G(2/4)125	G+121	G+122	G+123	G+124	
G+124	G*69	G+118	TN251		
G+123	G*67	G+117	TN250		
G+122	G*68	G+118	TN249		
G+121	G*66	G+117	TN248		
G+120	G+109	G+110			
G+119	G+107	G+108	TN252		
G+118	G*114	TN247			
G+117	G*113	TN246			
G*116	G+103	G+105			
G*115	G+102	G+104			
G*114	G+91	G+93			
G*113	G+90	G+92			
G(3/4)112	G+98	G+99	G+100	G+101	
G(3/4)111	G+94	G+95	G+96	G+97	
G(3/4)110	G+86 G+87	G+88	G+89		
G(3/4)109	G+82	G+83	G+84	G+85	
G(3/4)108	G+78	G+79	G+80	G+81	
G(3/4)107	G+74	G+75	G+76	G+77	
G(3/4)106	G+70	G+71	G+72	G+73	

G+105	G*69	TN245
G+104	G*67	TN244
G+103	G*68	TN243
G+102	G*66	TN242
G+101	G*69	TN241
G+100	G*67	TN240
G+99	G*68	TN239
G+98	G*66	TN238
G+97	G*69	TN237
G+96	G*67	TN236
G+95	G*68	TN235
G+94	G*66	TN234
G+93	G*69	TN233
G+92	G*67	TN232
G+91	G*68	TN231
G+90	G*66	TN230
G+89	G*65	TN229
G+88	G*64	TN228
G+87	G*63	TN227
G+86	G*62	TN226
G+85	G*65	TN225
G+84	G*64	TN224
G+83	G*63	TN223
G+82	G*62	TN222
G+81	G*65	TN221
G+80	G*64	TN220
G+79	G*63	TN219
G+78	G*62	TN218
G+77	G*65	TN217
G+76	G*64	TN216
G+75	G*63	TN215
G+74	G*62	TN214
G+73	G*65	TN213
G+72	G*64	TN212
G+71	G*63	TN211
G+70	G*62	TN210
G*69	TN201	TN209
G*68	TN201	TN208
G*67	TN201	TN207
G*66	TN201	TN206
G*65	TN201	TN205
G*64	TN201	TN204
G*63	TN201	TN203
G*62	TN201	TN202

B2.2 *Probabilistic data*

The following table gives the probabilities of occurrence for the basic events of the fault tree Europe-1.

TN201	.01	TN231	.0137
TN202	.051	TN232	.0137
TN203	.051	TN233	.0137
TN204	.051	TN234	.016
TN205	.051	TN235	.016
TN206	.112	TN236	.016
TN207	.112	TN237	.016
TN208	.112	TN238	.016
TN209	.112	TN239	.016
TN210	.016	TN240	.016
TN211	.016	TN241	.016
TN212	.016	TN242	.0038
TN213	.016	TN243	.0038
TN214	.0218	TN244	.0117
TN215	.0218	TN245	.0117
TN216	.0218	TN246	.00052
TN217	.0218	TN247	.00052
TN218	.015	TN248	.018
TN219	.015	TN249	.018
TN220	.015	TN250	.018
TN221	.015	TN251	.018
TN222	.016	TN252	.000008
TN223	.016	TN253	.000072
TN224	.016	TN254	.015
TN225	.016	TN255	.015
TN226	.015	TN256	.015
TN227	.015	TN257	.015
TN228	.015	TN258	.0188
TN229	.015	TN259	.0188
TN230	.0137	TN260	.0188
		TN261	.0188

B2.3 *Results*

$P_S = 1.282 \cdot 10^{-6}$
46188 minimal cut sets

B3 Fault tree: Europe-2

B3.1 *Structure of the fault tree*

G(3/5)1	G+71	G+70	G+69	G+68	G+67
G+71	G*66	TN32			
G+70	G*65	TN31			
G+69	G*64	TN30			
G+68	G*63	TN29			
G+67	G*62	TN28			
G*66	G+61	G+56			
G*65	G+60	G+56			
G*64	G+59	G+56			
G*63	G+58	G+56			
G*62	G+57	G+56			
G+61	G+55	TN27			
G+60	G+55	TN26			
G+59	G+55	TN25			
G+58	G+55	TN24			
G+57	G+55	TN23			
G+56	G+54	TN22			
G(2/3)55	G+53	G+52	G+51		
G(2/3)54	G+40	G+37	G+34		
G+53	G*47	TN21			
G+52	G*46	TN20			
G+51	G*45	TN19			
G+50	G*47	TN18			
G+49	G*46	TN17			
G+48	G*45	TN16			
G*47	G+44	TN15			
G*46	G+43	TN14			
G*45	G+42	TN13			
G(2/3)44	G+42	G+38	G+35		
G(2/3)43	G+40	G+37	G+34		
G(2/3)42	G+39	G+36	G+33		
G+41	TN12	TN10			
G+40	TN11	TN10			
G+39	TN9	TN10			
G+38	TN8	TN6			
G+37	TN7	TN6			
G*36	TN5	TN6			
G*35	TN4	TN2			
G*34	TN3	TN2			
G*33	TN1	TN2			

B3.2 *Probabilistic data*

The following table gives the probabilities of occurrence of the basic events of the fault tree Europe-2.

TN1	0.1
TN2	0.035
TN3	0.1
TN4	0.1
TN5	0.1
TN6	0.035
TN7	0.1
TN8	0.1
TN9	0.1
TN10	0.035
TN11	0.1
TN12	0.1
TN13	0.022
TN14	0.022
TN15	0.022
TN16	0.00088
TN17	0.00088
TN18	0.00088
TN19	0.00088
TN20	0.00088
TN21	0.00088
TN22	0.00175
TN23	0.00175
TN24	0.00175
TN25	0.00175
TN26	0.00175
TN27	0.00175
TN28	0.00875
TN29	0.00875
TN30	0.00875
TN31	0.00875
TN32	0.00875

B3.3 *Results*

$P_S = 7.822 \cdot 10^{-6}$
3412 minimal cut sets

B4 Fault tree: Europe-3

B4.1 *Structure of the FT*

G+187	G*183	G*175	G*186	
G*186	G+185	G+184		
G+185	G*155	G*141		
G+184	G*156	G*140		
G*183	G+182	G+181		
G+182	G*180	G+133	TN003	
G+181	G*179	G+132	TN001	
G*180	G*178	TC080		
G*179	G*178	TC079		
G*178	G+177	G+176		
G+177	G+120	TC078		
G+176	G*108	G*107	TC077	
G*175	G+174	G+173		
G+174	G*172	G+166		
G+173	G*171	G+165		
G*172	G+170	G+169		
G*171	G+168	G+167		
G+170	G+120	G+101	TC076	
G+169	G*108	G*107	G+100	TC075
G+168	G+119	G+094	TC074	
G+167	G*108	G*107	G+093	TC073
G+166	G*164	G*155		
G+165	G*163	G*156		
G*164	G*162	G+159		
G*163	G*161	G+157		
G*162	G+160	TN072		
G*161	G+158	TN071		
G+160	G+120	TC068	TC070	
G+159	G*108	G*107	TC068	TC069
G+158	G+119	TC065	TC067	
G+157	G*108	G*107	TC065	TC066
G*156	G+153	G*149		
G*155	G+154	G*148		
G+154	G*152	G*149		
G+153	G*152	G*148		
G*152	G+151	G+150		

G+151	G*108	G*107	TC064		
G+150	G+119	TC063			
G*149	G*147	G+144			
G*148	G*146	G+142			
G*147	G+145	TN062			
G*146	G+143	TN061			
G+145	G+096	TC060	TC059		
G+144	G+095	TC058	TC059		
G+143	G+101	TC057	TC056		
G+142	G+100	TC055	TC056		
G*141	G*139	G+136			
G*140	G*138	G+134			
G*139	G+137	TN054			
G*138	G+135	TN053			
G+137	G+120	TC052	TC051		
G+136	G*108	G*107	TC050	TC051	
G+135	G+120	TC049	TC048		
G+134	G*108	G*107	TC047	TC048	
G+133	G+131	TC046	TC035		
G+132	G+130	TC045	TC037		
G+131	G*118	G*129	TC044		
G+130	G*118	G*128	TC043		
G*129	G+127	G+124			
G*128	G+126	G+125			
G+127	G+125	TC042			
G+126	G+124	TC042			
G+125	G*123	G+094	TC041		
G+124	G*122	G+101	TC040		
G*123	G+121	TC039			
G*122	G+121	TC038			
G+121	G+101	G+094	TC035	TC036	TC037
G+120	G*118	G*117	TC034		
G+119	G*118	G*116	TC033		
G*118	G+101	G+094			
G*117	G+115	G+112			
G*116	G+114	G+113			
G+115	G+113	TC032			
G+114	G+112	TC032			
G+113	G*111	G+094	TC031		
G+112	G*110	G+101	TC030		

G*111	G+109	TC029			
G*110	G+109	TC028			
G+109	G+101	G+094	TC027		
G*108	G+100	G+093			
G*107	G+106	G+105			
G+106	G*104	G+093	TC026		
G+105	G*103	G+100	TC025		
G*104	G+102	TC024			
G*103	G+102	TC023			
G+102	G+100	G+093	TC022		
G+101	G*099	G*097	TC021		
G+100	G*098	G*097	TC020		
G*099	G+089	G+081			
G*098	G+090	TN019			
G*097	G+081	TN019			
G+096	G+094	TC010	TC018		
G+095	G+093	TC007	TC017		
G+094	G*092	G*088	G*087	G*082	TC016
G+093	G*086	G*085	G*091	G*082	TC016
G*092	G+083	TN004			
G*091	G+084	TN005			
G+090	TC014	TC015	TC013		
G+089	TC011	TC012	TC013		
G*088	TC009	TC010			
G*087	TC008	TC010			
G*086	TC008	TC007			
G*085	TC006	TC007			
G+084	TC009	TC010	TC008		
G+083	TC006	TC007	TC008		
G*082	TN004	TN005			
G+081	TN001	TN002	TN003		

B4.2 *Probabilistic data*

All the components (TN and TC) have a rate of appearance equal to $10^{-6}h^{-1}$

B4.3 *Results*

24386 minimal cut sets

Appendix C

Some Results of Probabilities

This appendix contains the elements of the theory of probabilities that are necessary in the statement of the Monte Carlo method.

Proposition C.1 (Chebychev inequality)
Consider a square intergrable r.v X. Then for all $\varepsilon > 0$, we have:

$$P(|X - \mathbb{E}X| \geq \varepsilon) \leq \frac{Var(X)}{\varepsilon^2} \tag{C.1}$$

Definition C.2 (i.i.d sequences)
A sequence of r.v $(X_n, n \geq 1)$ defined on the probability space $(\Omega, \mathcal{F}, \mathbb{P})$ is said to be i.i.d, when any finite sequence extracted from this sequence is of the i.i.d type Set $S_n = X_1 + \ldots + X_n . n \geq 1$.

Proposition C.3 (Week law of large numbers: Khintchine theorem)
Let there be a sequence of r.v $(X_n, n \geq 1)$ i.i.d with $\mathbb{E}|X_1| < \infty$. When $n \to \infty$, we have

$$\frac{S_n}{n} \xrightarrow{p} \mathbb{E}X_1 \tag{C.2}$$

Proposition C.4 (Strong law of large numbers: Kolmogorov theorem)
Let there be a sequence of r.v $(X_n, n \geq 1)$ *i.i.d. Then,* $\mathbb{E}|X_1| < \infty$ *is a necessary and sufficient condition such that the sequence* (X_n) *verifies the law of large numbers, i.e. when* $n \to \infty$*, we have*

$$\frac{S_n}{n} \xrightarrow{p.s.} \mathbb{E}X_1 \tag{C.3}$$

▷ **Example C.3.** For a series of independent events $(A_n, n \geq 1)$ and of the same probability p, the strong law of large numbers says that the frequencies $\frac{1}{n}\sum_{k=1}^{n} 1_{A_k}$ converge almost surely (a.s) towards p when $n \to \infty$. This result justifies the estimation of the probabilities through the frequencies.

▷ **Example C.4.** The function of empirical distribution of an n-sample is defined by

$$F_n(x) = \frac{1}{n}\sum_{i=1}^{n} 1_{\{X_i \leq x\}} \tag{C.4}$$

From the strong law of large numbers, we directly obtain, for any $x \in \mathbb{R}$, and when $n \to \infty$,

$$F_n(x) \xrightarrow{a.s.} F(x) \tag{C.5}$$

Proposition C.5 (Central limit theorem (CLT))
Let there be a sequence of r.v $(X_n, n \geq 1)$*, i.i.d with common expectation* μ *and common variance* σ^2 *such that* $0 < \sigma^2 < \infty$*. Then, when* $n \to \infty$*, we have*

$$\frac{S_n - n\mu}{\sigma\sqrt{n}} \xrightarrow{\mathcal{L}} N(0,1) \tag{C.6}$$

▷ **Example C.5.** (Moivre-Laplace theorem). Let $(X_n, n \geq 1)$ be a sequence of r.v i.i.d with $X_n \sim \mathcal{B}(p), (0 < p < 1)$, then $\mu = p$ and $\sigma^2 = p(1-p) > 0$. According to proposition C.5, we have

$$\frac{S_n - np}{\sqrt{np(1-p)}} \xrightarrow{\mathcal{L}} N(0,1). \tag{C.7}$$

▷ **Example C.6.** The application of the CLT to the sequence of empirical functions $F_n(\cdot)$, when $n \to \infty$, gives, for all $x \in \mathbb{R}$

$$\sqrt{n}(F_n(x) - F(x)) \xrightarrow{\mathcal{L}} N(0, \sigma^2(x)) \tag{C.8}$$

with $\sigma^2(x) = F(x)(1 - F(x))$.

Proposition C.6 (CLT for the r.v vector)

Let there be a sequence of random vectors $(X_n, n \geq 1)$, i.i.d with values in $\mathbb{R}^d$ of mean vector $\mu \in \mathbb{R}^d$, and common variances-covariances matrices K. Then, when $n \to \infty$, we have

$$\frac{1}{\sqrt{n}}(S_n - n\mu) \xrightarrow{\mathcal{L}} N_d(0, K). \tag{C.9}$$

where $N_d(0, K)$ is the d-dimensional normal distribution, with mean the vector $0 \in \mathbb{R}_d$ and variance-covariances matrix K.

Proposition C.7 (Law of the iterated logarithm (LIL))

Let there be a sequence of real r.v $(X_n, n \geq 1)$, i.i.d with $\mathbb{E}X_1 = 0$ and $Var(X_1) = 1$ Then

$$\overline{\lim_{n \to \infty}} \frac{S_n}{\sqrt{2n \log \log n}} = 1, \quad a.s \tag{C.10}$$

$$\underline{\lim_{n \to \infty}} \frac{S_n}{\sqrt{2n \log \log n}} = -1, \quad a.s \tag{C.11}$$

Example C.6. The application of the CLT to the sequence of empirical functions $F_n(x)$, when $n \to \infty$, gives for all $x \in \mathbb{R}$

$$\sqrt{n}(F_n(x) - F(x)) \xrightarrow{L} N(0, \sigma^2(x)) \quad (C.8)$$

with $\sigma^2(x) = F(x)(1 - F(x))$.

Proposition C.6 (CLT for the r.v. vectors)

Let [illegible] *be a sequence of random vectors* [illegible] *of mean vector* $\mu \in \mathbb{R}^d$ [illegible] *matrices* Σ. [illegible]

[illegible]

$$[illegible] \; N_d(0, \Lambda) \quad (C.9)$$

where $N_d(0, \Lambda)$ *is the* [illegible] *normal distribution* [illegible]

[illegible]

Proposition C.7 (Law of the iterated logarithm (LIL))

[illegible]

Then

$$\overline{\lim_{n \to \infty}} \frac{S_n}{\sqrt{2n \log \log n}} = 1, \quad [illegible] \quad (C.10)$$

$$[illegible] \lim \frac{[illegible]}{\sqrt{2n \log \log n}} = -1, \quad a.s. \quad (C.11)$$

Main Notations

F	c.d.f of a random variable (r.v)
f	p.d.f of an r.v
$R(t)$	rcliability at time t
$A(t)$	instantaneous availability
$M(t)$	maintainability
Q, P_S	probability of the top event of a fault tree
A_∞ or A	asymptotic availability
$\widetilde{A}(t)$	average availability over $[0,t]$
$\widetilde{A}_\infty$ or $\widetilde{A}$	asymptotic average availability
$h(t)$	hazard rate at time t
$\lambda(t)$	instantaneous failure rate
$\mu(t)$	instantaneous repair rate
$L(t)$	mean residual lifetime at time t
T	lifetime of the system
θ	duration of mission of the system
Y	duration of the repair
$\mathbf{1}_A$	indicator function of event A
m_k	k-th centered moment of the random variable X $(k \in \mathbb{N}^*)$
$\Gamma(x)$	Gamma function: $\Gamma(x) = \int_0^\infty t^{x-1}e^{-t}dt$. if $x \in \mathbb{N}^*$ then $\Gamma(x) = (x-1)!$
$\mathbb{R}$	set of real numbers: $\mathbb{R} = (-\infty, +\infty)$ and $\mathbb{R}_+ = [0, +\infty)$
$\mathbb{N}$	set of natural numbers: $\mathbb{N} = \{0, 1, 2, ...\}$ and $\mathbb{N}^* = \mathbb{N} \setminus \{0\}$
φ	structure function of a system
C	set of components of a system
$\mathcal{E}$	state space of a system

$\mathcal{E}_i$	state space of a component $i \in C$
K_i	i-th minimal cut set of a system or of an FT
C_i	i-th minimal path of a system or of an FT
$\mathcal{K}$	set of minimal cut sets of a system or of an FT
$\mathcal{C}$	set of minimal paths of a system or of an FT
$\mathbf{x}$	vector of states of the components of the system, $\mathbf{x} = (x_1, ..., x_n)$, $(a_i, \mathbf{x})$ vector $\mathbf{x}$ with its i-th element fixed at a, $a = 0$ or 1
$\mathbf{X}$	random vector of the states of the components of the system, $\mathbf{X} = (X_1, ..., X_n)$
EX	expectation of the r.v. X
$\overline{A}$	contrary event, i.e., $\overline{A} = \Omega \setminus A$
$\cup$	disjunction of two events
$\cap$	conjunction of two events

Abbreviations

FT	fault tree
c-FT	coherent fault tree
nc-FT	non-coherent fault tree
FT-r	fault tree with restrictions
FT-d	fault tree with delay
m-FT	multistate fault tree
BDD	binary decision diagram
p.d.f	probability density function
c.d.f	cumulative distribution function
i.i.d	independent and identically distributed
MC	common mode of failure
MTTF	mean time to failure
MTTR	mean time to repair
MUT	mean up-time
MDT	mean down-time
MTBF	mean time between failures
r.v	random variable

Conventions

k-out-of-n: G	a system whose good functioning is obtained through the good functioning at k out of n components
k-out-of-n: F	a system whose failure is obtained by the failure of at least k out of n components
$a - b$	the components a and b are in series
$a//b$	the components a and b are in parallel (active redundancy)
a/b	the components a and b are in parallel (passive redundancy with a as active component and b as standby)

Bibliography

[**ABR 79**] J.A. Abraham, "An improved algorithm for network reliability", IEEE Trans. on Reliab., vol. R-28, N°1, April 1979, pp. 58–61.

[**ALL 83**] D.J. Allen, "Digraphs and fault trees", Hazard Prevention, Jan./Febr. 1983, pp. 22–25.

[**AKE 87**] S.B. Akers, "Binary decision diagrams", IEEE Trans. Computers, vol. C-27, N°6, 1978, pp 509–516.

[**AME 85**] A. Amendola, "Reliability data bases", Reindel P.C., Dordrecht-Hollande, 1985.

[**AME 86**] H.H. Amer, R.K. Iyer, "Effect of uncertainty in failure rates on memory systems reliability", IEEE Trans. on Reliab., vol. R-35, N°4, Oct. 1986, pp. 377–379.

[**AND 80**] P.K. Androw, "Difficulties in fault-tree synthesis for process plant", IEEE Trans. on Reliab., vol. R-29, N°1, April 1980, pp. 2–9.

[**APO 76**] G. Apostolakis, D. Okren, S.L. Salem, "A computer-oriented app.roch to fault-tree construction", EPRI NP 288, Univ. of California, Nov. 1976.

[**ANG 94**] J.E. Angus, "The probability integral transform and related results", SIAM Review, vol. 36, N°4, pp. 652-654.

[**APO 77**] G. Apostolakis, Y.T. Lee, "Methods for the estimation of confidence bounds for the top-event unavailability of fault trees", Nucl. Eng. Des., vol. 41, 1977, pp. 411–419.

[**APO 81**] G. Apostolakis, S. Kaplan, "Pitfalls in risk calculations", Reliab. Eng., vol. 2, 1981, pp. 135–145.

[**APO 87**] G. Apostolakis, P. Moieni, "The foundations of models of dependance in probabilistic safety assessment", Reliab. Eng., vol. 18, N°3, 1987, pp. 177–196.

[**ASC 84**] H.E. Ascher, H. Feingold, Repairable systems reliability: Modeling, Inference, Misconception and their Causes, Marcel Dekker, 1984, NY.

[**AST 80**] M. Astolfi, S. Contini, C.L. Van den Muyzenberg, G. Volta, "Fault tree analysis by list-processing techniques", in Synthesis and analysis methods for safety and reliability studies, In G. Apostolakis, S. Garribba, G. Volta (eds.), Plenum Press, N.Y., 1980, pp. 5–32.

[**AVE 86**] T. Aven, "On the computation of certain measures of importance of system components", Microelectron. Reliab., vol. 26, N°2, 1986, pp. 279–281.

[**AVE 86**] T. Aven, R. Osteno, "Two newcomponent importance measures for a flow network system", Reliab. Eng., vol. 4, 1986, pp. 75–80.

[**BAR 75**] R.E. Barlow, F. Proschan, "Importance of system components and fault tree events", Stoch. Proc. & Appl., vol. 3, 1975, pp. 153–173.

[**BAR 75**] R.E. Barlow, F. Proschan , Statistical Theory of Reliability and Life Testing- Probability Models, Holt, R-W, Inc., 1975.

[**BAR 78**] R.E. Barlow, A.S. Wu, "Coherent systems with multi-state components", Math. Oper. Res., vol. 3, N°4, Nov. 1978, pp. 275–281.

[**BAX 84**] L.A. Baxter, "Continuum structure I", J. of Appl. Probab., vol. 19, 1984, pp. 391–402.

[**BEN 85**] A. Bendell, L.A. Walls, "Exploring reliability data", Qual. Reliab. Eng. Intern., vol. 1, 1985, pp. 37–51.

[**BEN 76**] N.N. Bengiamin, B.A. Bowen, K.F. Schenk, "An efficient algorithm for reducing the complexity of computation in fault tree analysis", IEEE Trans. Nucl. Sci., vol. NS-23, N°5, Oct. 1976, pp. 1442–1446.

[**BEN 75**] R.G. Bennetts, "On the analysis of fault trees", IEEE Trans. on Reliab., vol. R-24, N°3, August 1975, pp. 175–185.

[**BIE 83**] V. Bier, "A measure of uncertainty importance for components in fault trees", PhD Thesis, MIT, LIDS-TH-1277, Jan. 1983, Cambridge.

[**BIR 65**] Z.W. Birnbaum, J.D. Esary, "Modules of coherent binary systems", J. Soc. Indust. Appl. Math., vol. 13, N°2, June 1965, pp. 444–462.

[**BIR 69**] Z.W. Birnbaum, "On the importance of different components in a multicomponent system", in Multivariate Analysis II, P. Krishnaiam (ed), Academic Press, 1969, pp. 581–592.

[**BLI 80**] A. Blin, A. Carnino, J.P. Signoret, F. Buscatie, B. Duchemin, J.M. Landré, H. Kalli, and M.J. De Villeneuve, "PATREC, a computer code for fault-tree calculation", in [ASTOL 80], pp. 33–44.

[**BLO 82**] H.W. Block, T.H. Savits, "A decomposition for multistate monotone systems", J. Appl. Prob., vol. 19, 1982, pp. 391–402.

[**BON 96**] J.L Bon, Fiabilité des systèmes, 1996, Dunod, Paris.

[**BOS 84**] A. Bossche, "The top-event's failure frequency for non-coherent multi- state fault trees", Microelectron. Reliab., vol. 24, N°4, 1984, pp. 707–715.

[**BOS 87**] A. Bossche, "Calculation of critical importance for multi-state components", IEEE Trans. on Reliab., vol. R-36, N°2, June, 1987, pp. 247–249.

[**BOU 91**] M. Bouissou, "Une heuristique d'ordonnancement des variables pour la construction des diagrammes de décision", 9th National Conference $\lambda\mu$, 1994, pp 513–518, La Baule.

[**BOU 86**] N. Bouleau, Probabilités de l'ingénieur, variables aléatoires et simulation, Hermann, 1986.

[**BRO 90**] K.S. Brown, "Evaluating fault tree (AND & OR gates only) with repeated events", IEEE Trans. on Reliab., vol. 39, N°2, Dec. 1990, pp. 226–235.

[**BRA 03**] C. Bracquemond, O. Gaudoin "A survey on discrete lifetime distributions", Quality Safety Eng., vol. 10, No 1, 2003, pp. 69–98.

[**BRY 87**] R.E. Bryant, "Graph based algorithms for boolean function manipulation", IEEE Trans. Computer, vol. 35, N°8, 1987, pp 677–691.

[**BRY 92**] R.E. Bryant, "Symbolic boolean manipulation with ordered binary-decision diagrams", ACM Computing Surveys, vol. 24, N°3, Sept. 1992, pp 293–318.

[**BUT 77**] D.A. Butler, "An importance ranking for system components based upon cuts", Oper. Research, vol. 25, N°5, 1977, pp. 874–879.

[**BUT 79**] D.A. Butler, "A complete importance ranking for components of binary coherent systems, with extentions to multi-state systems", Naval Res. Log. Quart., vol. 26, 1979, pp. 565–578.

[**BUZ 83**] J.A. Buzacott, "The ordering of terms in cut-based recursive disjoint products", IEEE Trans. on Reliab., vol. R-32, N°5, Dec. 1983, pp. 472–474.

[**BUZ 83**] J.A. Buzacott, G.J. Anders, "Probability of component or subsystem failure before system failure", IEEE Trans. on Reliab., vol. R-32, N°5, Dec. 1983, pp. 450–452.

[**CAL 80**] L. Caldarola, "Unavailability and failure intensity of components", Nucl. Eng. Des. J., 44, 1997, pp 147.

[**CAL 80**] L. Caldarola, "Coherent systems with multistate components", Nucl. Eng. and Des., vol. 58, 1980, pp. 127–139.

[**CAM 78**] P. Camarda, F. Corsi, A. Trentadue, "An efficient simple algorithm for fault tree automatic systhesis from the reliability graph", IEEE Trans. on Reliab., vol. R-27, N°3, Aug. 1978, pp. 215–221.

[**CAR 86**] G. Carpinelli, U. De Martinis, A. Losi, V. Mangoni, R. Mongelluzzo, A. Piccolo, A. Testa, V. Vaccaro, "Some considerations on computer-aided fault-tree construction for electric power systems components", Technical note N°1.05.E3.86.41, PER 1086/86, May 1986.

[**CHA 86**] K.C. Chae, "System reliability in the presence of common-cause failures", IEEE Trans. on Reliab., vol. R-35, N°1, April 1980, pp. 32–35.

[**CHA 74**] P. Chatterjee, "Fault tree analysis: Reliability theory and systems safety analysis", Oper. Res. Center, University of California, Berkeley, 1974.

[**CHA 75**] P. Chatterjee, "Modularization of fault trees: A method to reduce the cost of analysis", in Reliability and Fault Tree Analysis. Theoretical and applied aspects of system reliability and safety assessment, R.E. Barlow, J.B. Fussell, N.D. Singpurwalla (eds), Siam, Philadelphia, 1975, pp. 101–126.

[**CHU 80**] T.L. Chu, G. Apostolakis, "Methods for probabilistic analysis of noncoherent fault trees", IEEE Trans. on Reliab., vol. R-39, N°5, Dec. 1980, pp. 354–360.

[**CLA 81**] C.A. Clarotti, "Limitations of minimal cut-set approach in evaluating reliability of systems with repairable components", IEEE Trans. on Reliab., vol. R-30, N°4, Oct. 1981, pp. 335–338.

[**CLA 81**] C.A. Clarotti, F. Spizzichino, "Minimal-cut reliability lower-bound for systems containing standby modules", IEEE Trans. on Reliab., vol. R-30, N°3, Aug. 1981, pp. 293–297.

[**COC 97**] C. Cocozza-Thivent, Processus stochastiques et fiabilité des systèmes, SMAI, Springer, 1997.

[**COL 80**] A.G. Colombo, "Uncertainty propagation in fault-tree analysis", in Synthesis and analysis methods for safety and reliability studies, G. Apostolakis, S. Garribba, G. Volta (eds.), Plenum Press, N.Y., 1980, pp. 95–103.

[**COR 75**] M. Corazza, Techniques Mathématiques de la Fiabilité Prévisionnelle, Sup'Aero, Cependus, 1975.

[**COU 92**] O. Coudert, J.C. Madre, "A new method to compute prime and essential prime implicants of boolean functions", in Advanced Research in VLSI and Parallel Systems, Knight and Savage (Eds), pp 113–128. The MIT Press, March 1992.

[**COU 94**] O. Coudert, J.C. Madre, "MetaPrime: an interactive fault-tree analyzer", IEEE Trans. Reliability, vol. 43, N°1, March 1994, pp 121–127.

[**COX 82**] D.C. Cox, "An analytic method for uncertainty analysis of nonlinear output functions, with applications to fault-tree analysis", IEEE Trans. on Reliab., vol. R-31, N°5, Oct. 1982, pp. 465–469.

[**COX 77**] N.D. Cox, "Comparison of two uncertainty analysis methods", Nucl. Science and Eng, vol. 64, 1977, pp. 258–265.

[**DES 83**] W. De Souza Borges, F.W. Rodrigues, "An axiomatic characterization of multistate coherent structures", Math. Oper. Res., vol. 8, N°3, Aug. 1983, pp. 435–438.

[**DHI 86**] B.S. Dhillon, Human Reliability with Human Factors, Pergamon Press, 1986, Oxford.

[**DUB 80**] B. Dubuisson, M. Hassan, "A simplified method for evaluating time dependent fault-trees using Weibull distribution", Microelectron. Reliab., vol. 20, 1980, pp. 347–350.

[**DUG 91**] J.B. Dugan, "Automated analysis of phased-mission reliability", IEEE Trans. on Reliab., vol. 40, N°1, April, 1991 pp. 45–52.

[**EL-N 78**] E. El-Neweihi, F. Proschan, J. Sethuraman, "Multistate coherent systems", J. Appl. Prob., vol. 15, 1978, pp. 675–688.

[**ESA 75**] J.D. Esary, H. Ziehms, "Reliability analysis of phased systems", in [CHAT 75], pp. 213–236.

[**FLU 90**] B.D. Flury, "Acceptance-rejection sampling made easy", SIAM Review, vol. 32, N°3, Sept. 1990, pp. 474–476.

[**FON 87**] C.C. Fong, J.A. Buzacott, "Improved bounds for system-failure probability", IEEE Trans. on Reliab., vol. R-36, N°4, Oct. 1987, pp. 454–458.

[**FUS 72**] J.B. Fussell, W.E. Vesely, "A new methodology for obtaining cut sets", Amer. Nucl. Soc. Trans., vol. 15, N°1, June 1972, pp. 262–263.

[**FUS 76**] J.B. Fussell, E.F. Aber, R.G. Rahl, "On the quantitative analysis of priority-AND failure logic", IEEE Trans. on Reliab., vol. R-25, N°5, Dec. 1976, pp. 324–326.

[**GAR 77**] S. Garriba, P. Mussio, F. Naldi, G. Reina, G. Volta, "Efficient construction of minimal cut sets from fault trees", IEEE Trans. on Reliab., vol. R-26, N°2, June 1977, pp. 88–94.

[**GAR 85**] S. Garribba, E. Guagnini, P. Mussion, "Multiple-valued logic trees: meaning and prime implicants", IEEE Trans. Reliability, vol. R-34, N°5, Dec. 1985, pp. 463–472.

[**GIR 01**] V. Girardin, N. Limnios, Probabilités en vue des applications, Vuibert, 2001, Paris.

[**GNE 72**] B. Gnedenko, Y. Beliaev, A. Soloviev, Méthodes Mathématiques en Théorie de la Fiabilité, MIR, Moscou, 1972.

[**GON 78**] G.E. Gonzalez-Urdaneta, B.J. Cory, "Variance and approximate confidence limits for probability and frequency of system failure", IEEE Trans. on Reliab., vol. R-27, N°4, Oct. 1978, pp. 289–293.

[**GRE 96**] E.P. Greenwood, M.S. Nikulin, A guide to chi-squared testing, Wiley, N.Y., 1996.

[**GRI 80**] W.S. Griffith, "Multistate reliability models", J. Appl. Prob., vol. 17, 1980, pp. 735–744.

[**GRI 82**] W.S. Griffith, "A multistate availability model: system performance and component importance", IEEE Trans. Reliability, vol. R-31, N°1, April 1982, pp. 97–98.

[**HEI 82**] K.D. Heidtmann, "Bounds on reliability of noncoherent system using its length & width", IEEE Trans. on Reliab., vol. R-31, N°5, Dec. 1982, pp. 424–427.

[**HEI 83**] K.D. Heidtmann, "Inverting paths & cuts of 2-state systems", IEEE Trans. on Reliab., vol. R-32, N°5, Dec. 1983, pp. 469–471 & 474.

[**HEI 84**] K.D. Heidtmann, "A prior error estimates for the method of inclusion-exclusion with applications", Siam J. Appl. Math., vol. 44, N°2, April 1984, pp. 443–450.

[**HEN 81**] E.J. Henley, H. Kumamoto, Reliability Engineering and Risk Assessment, Prentice-Hall, 1981.

[**HIL 90**] C. Hill, C. Com-Nougué, A. Kramar, T. Moreau, J. O'Quigley, R. Senoussi, and C. Chastang "Analyse statistique des données de survie", Flamarion, Paris, 1990.

[**HOU 65**] R.W. House, T. Rado, "A generalization of Nelson's algorithm for obtaining prime implicants", The J. of Symb. Logic, vol. 30, N°1, March 1965, pp. 8–12.

[**HUD 83**] J.C. Hudson, K.C. Kapur, "Models in coherent multistate systems", IEEE Trans. Reliability, vol. R-32, N°2, June 1983, pp. 183–185.

[**HUG 87**] R.P. Hughes, "Fault tree truncation error bounds", Reliab. Eng., vol. 17, 1987, pp. 37–46.

[**HUL 75**] B.L. Hulme, R.B. Worrell, "A prime implicant algorithm with factoring", IEEE Trans. on Computers, Nov. 1975, pp. 1130–1131.

[**INA 80**] T. Inagaki, E.J. Henley, "Probabilistic of prime implicants and top-events for non coherent systems", IEEE Trans. on Reliab., vol. R-29, N°5, Dec.1980, pp. 361–67.

[**ION 99**] D.C. Ionescu, N. Limnios (Eds) , Statistical and Probabilistic Models in Reliability, Birkhauser, Boston, 1999.

[**JAC 82**] P.S. Jackson, "A second-order moments method for uncertainty analysis", IEEE Trans. on Reliab., vol. R-31, N°4, Oct. 1982, pp. 382–384.

[**JAC 83**] P.S. Jackson, "On the s-importance of elements and prime implicants of non-coherent systems", IEEE Trans. on Reliab., vol. R-32, N°1, April 1983, pp. 21–25.

[**JAN 85**] X. Janan, "On multistate system analysis", IEEE Trans. Reliability, vol. R-34, N°4, Oct. 1985, pp. 329–337.

[**JAN 99**] J. Janssen, N. Limnios (Eds), Semi-Markov processes and applications, Kluwer, Dordrecht, 1999.

[**JEA 85**] J.P. Jeannette, N. Limnios, "Méthodes actuelles d'analyse de sûreté des systèmes", R.G.S., N°43, Avril 1985, pp. 46–52.

[**JEA 86**] J.P. Jeannette, N. Limnios, "Contribution of safety analysis methods in the framework of safety audits", REL-CON'86, Eur. Rel. Conf., 16–20 June 1986, Copenhagen. In Reliability & Technology: Theory and Applications, F. Jensen and J. Moltoft (eds), North-Holland Publ. Co, 1986.

[**KAU 75**] A. Kaufmann, D. Grouchko, R. Cruon, Modèles Mathématiques pour l'Etude de la Fiabilité des Systèmes, Masson et Cie, Paris, 1975.

[**KER 91**] F. Kervégant, N. Limnios, C. Cocozza-Thivent, A. Dubreuil-Chambardel, "Evaluation probabiliste approchée des grands arbres de défaillance", RAIRO, operational research, 1991.

[**KIS 83**] L. Kiss, "From fault-tree to fault-identification", IEEE Trans. on Reliab., vol. R-32, N°5, Dec. 1983, pp. 422–425.

[**KOH 89**] T. Kohda, E.J. Henley, K. Inoue, "Finding modules in fault tree", IEEE Trans. on Reliab., vol. 38, N°2, June 1989, pp. 165–176.

[**KUM 77**] H. Kumamoto, K. Tanaka, K. Inoue, "Efficient evaluation of system reliability by Monte Carlo method", IEEE Trans. on Reliab., vol. R-26, N°5, Dec. 1977, pp. 311–315.

[**KUM 78**] H. Kumamoto, E.J. Henley, "Top-down algorithm for obtaining prime implicant sets of non-coherent fault trees", IEEE Trans. on Reliab., vol. R-27, N°4, Oct. 1978, pp. 242–249.

[**LAI 86**] C.D. Lai, "Bounds on reliability of a coherent system with positively correlated components", IEEE Trans. on Reliab., vol. R-35, N°5, Dec. 1986, pp. 508–511.

[**LAP 77**] S.A. Lapp, G.J. Powers, "Computer-aided synthesis of fault-trees", IEEE Trans. on Reliab., April 1977, pp. 2–13.

[**LAP 79**] S.A. Lapp, G.J. Powers, "Update of Lapp-Powers fault-tree synthesis algorithm", IEEE Trans. on Reliab., vol. R-28, N°1, April 1979, pp. 12–14.

[**LEE 81**] K.K. Lee, "A compilation technique for exact system reliability", IEEE Trans. on Reliab., vol. R-30, N°3, Aug. 1981, pp. 335–338.

[**LEE 59**] C.Y. Lee, "Representation of switching circuits by binary-decision diagrams", Bell. Syst. Tech. J., vol. 38, July 1959, pp 985–999.

[**LIA 90**] H.T. Liaw, C.S. Lin, "On the OBDD-representation of general boolean functions", NSC Rep., N°NSC79–0404-E002–35, July 1990.

[**LEV 83**] N.G. Leveson, J.L. Stolzy, "Safety analysis of Ada programs using fault trees", IEEE Trans. on Reliab., vol. R-32, N°5, Dec.1983, pp. 479–484.

[**LIE 75**] C. Lievens, "Sécurité des systèmes", Sup'Aero, Cependus, Toulouse, 1975.

[**LIM 84**] N. Limnios, J.P. Jeannette, "Les logiciels AREV et ARBRE", 3e Sém. Europ.3SF, 19–21 Sept., 1984, Cannes.

[**LIM 85**] N. Limnios, J.P. Jeannette, "Traitement automatique des arbres de défaillance", Le Nouvel Automatisme, N°51, Jan.-Feb. 1985, pp. 56–62.

[**LIM 86**] N. Limnios, R.Ziani, "An algorithm for reducing cut sets in fault-tree analysis", IEEE Trans. on Reliab., vol. R-35, N°5, Dec. 1986, pp. 559–562.

[**LIM 86**] N. Limnios, R. Ziani, J.F. Guyonnet, "Traitement automatique des arbres de défaillance non cohérents. Le logiciel ARBRE-NC", 4th Sém. Eur., 3SF, 4–6 juin, 1986, Deauville.

[**LIM 87**] N. Limnios, "Failure rate of non-repairable systems", Reliability Engineering, vol. 17, N°2, 1987, pp. 83–88.

[**LIM 87**] N. Limnios, J.P. Jeannette, "Event trees and their treatment on PC computers", Reliability Engineering, vol. 18, N°3, 1987, pp. 197–204.

[**LIM 87**] N. Limnios, "A formal definition of fault-tree graph models and an exhaustive test of their stuctural data", Reliability Engineering, vol. 18, N°4, 1987, pp. 267–274.

[**LIM 90**] N. Limnios, "Failure-delay systems reliability modelling", In Systems Reliability Assessment, A.G. Colombo, A. Saiz de Boustamante (eds.), 1990 ECSC, EEC, EAEC, Brussels & Luxembourg, 1990, pp. 145–158.

[**LIM 96**] N. Limnios, A. Rauzy (Eds), Binary Decision Diagrams and Reliability", Spacial issue, Journal Eur. Systèmes Automatisés.

[**LIM 00**] N. Limnios, M. Nikulin, (Eds), Recent advances in reliability theory: Inference, methods and practice, Birkhauser, Boston, 2000.

[**LIM 01**] N. Limnios, G. Oprisan, Semi-Markov processes and reliability, Birkhauser, Boston, 2000.

[**LOC 78**] M.O. Locks, "Relationship between minimal path sets and cut sets", IEEE Trans. on Reliab., vol. R-27, N°2, June 1978, p. 106.

[**LOC 78**] M.O. Locks, "Inverting and minimalizing path sets and cut sets", IEEE Trans. on Reliab., vol. R-27, N°2, June 1978, p 107–109.

[**LOC 79**] M.O. Locks, "Synthesis of fault trees: an example of noncoherence", IEEE Trans. on Reliab., vol. R-28, N°1, Dec. 1979, pp. 2–5.

[**LOC 81**] M.O. Locks, "Modularizing, minimizing, and interpreting the K&H fault tree", IEEE Trans. on Reliab., vol. R-30, N°5, Dec. 1981, pp. 411–415.

[**LU 86**] P.-E. Lu, D.-H. Shi, "A simple method for calculating the structural importance of a component", Microelectron. Reliab., vol. 26, N°5, 1986, pp. 857–858.

[**MAR 75**] J.D. Marchland, "Fundamental concepts and relations for reliability analysis of multi-state systems", in Reliability and Fault Tree Analysis, R.E. Barlow and J.B. Fussell (Eds), SIAM, 1975, Pennsylvania.

[**MAS 87**] M. Masera, "Uncertainty propagation in fault tree analyses using lognormal distributions", IEEE Trans. on Reliab., vol. R-36, N°1, April 1987, pp. 145–149.

[**MEL 83**] W.A. Metler, D. Dietrich, "A moments compiler for computing Bayes intervals for complex systems", IEEE Trans. on Reliab., vol. R-32, N°1, April 1983, pp. 111–114.

[**MIS 92**] K.B. Misra, Reliability Analysis and Prediction, Elsevier, 1992.

[**MOD 75**] M. Modarres, "Mathematical concepts of time-dependent modularized fault tree evaluation", in [ESAR 75], pp. 1257–1266.

[**MOD 79**] M. Modarres, "Reliability analysis of complex technical systems using the fault tree modularization technique", PhD Thesis, MIT Depart. of Nucl. Eng., 1979.

[**MOD 84**] M. Modarres, H. Dezfuli, "A truncation methodology for evaluating large fault trees", IEEE Trans. on Reliab., vol. R-33, N°4, Oct. 1984, pp. 325–328.

[**MON 90**] J. Montero, J. Tejada, J. Yànez, "Some mathematical properties of multistate continuum systems", 7th Coll. Intern. Fiabilité et Maintenabilité, 1990, Brest, France, pp. 82–87.

[**MOS 91**] A. Mosleh (Guest editor), "Dependent failure analysis", Reliability Eng. & Syst. Saf., vol. 34, N°3, 1991.

[**NAK 79**] K. Nakashima, Y. Hattori, "An efficient bottom-up algorithm for enumerating minimal cut sets of fault trees", IEEE Trans. on Reliab., vol. R-28, N°5, Dec. 1979, pp. 353–357.

[**NAK 82**] K. Nakashima, K. Yamato, "Variance-importance of system components", IEEE Trans. on Reliab., vol. R-31, N°1, April 1982, pp. 99–100.

[**NAT 79**] B. Natvig, "A suggestion of a new measure of importance of system components", Stoch. Proc. & Appl., vol. 9, 1979, pp. 319–330.

[**NAT 80**] B. Natvig, "Improved bounds for the availability and unavailability in a fixed time interval for systems of maintenaid, independent components", Adv. Appl. Prob., vol 12, 1980, pp. 200–221.

[**NAT 82**] B. Natvig, "On the reduction in remaing system lifetime due to the failure of a specific component", J. Appl. Prob., vol. 19, 1982, pp. 642–652.

[**NAT 82**] B. Natvig, "Two suggestions of how to define a multistate coherent system", Adv. Appl. Prob., vol. 14, 1982, pp. 434–455.

[NEL 54] R.J. Nelson, “Simplest normal truth functions”, The J. of Symb. Logic, vol. 20, N°2, June 1954, pp. 105–108.

[NEL 55] R.J. Nelson, “Weak simplest normal truth functions”, The J. of Symb. Logic, vol. 20, N°3, Sept. 1955, pp. 232–234.

[OGU 81] E.I. Ogunbiyi, E.J. Henley, “Irredundant forms and prime implicants of a function with multistate variables”, IEEE Trans. on Reliab., vol. R-30, N°1, April 1981, pp. 39–42.

[ODE 95] K. Odeh, “Nouveaux algorithmes pour le traitement probabiliste et logique des arbres de défaillance”, PhD Thesis, University of Technology. Compiègne, Dec. 1995.

[ODE 93] K. Odeh, N. Limnios, “An efficient method for probability evaluation of a fault tree”, In System Modeling and Optimization, J. Henry & J.P.Yvon (Eds), pp. 951–957, 1993, Springer-Verlag.

[ODE 94] K. Odeh, N. Limnios, “Comparaison de principales méthodes d’évaluation probabiliste des arbres de défaillance”, 9th National Conference $\lambda\mu - 9$, 1994, pp 513–518, La Baule.

[ODE 96] K. Odeh, N. Limnios, “A new algorithm for fault-trees prime implicants computation”, International conference, ESREL’96, Crete (Greece), June 1996.

[OHI 84] F. Ohi, T. Nishida, “On multistate coherent systems”, IEEE Trans. Reliability, vol. R-33, N°4, Oct. 1984, pp. 284–287.

[OLM 78] J. Olmos, L. Wolf, “A modular representation and analysis of fault trees”, Nucl. Eng. Des., vol. 48, 1978, pp. 531–561.

[PAG 86] L.B. Page, J.E. Perry, “An algorithm for exact fault-tree probabilities without cut sets”, IEEE Trans. on Reliab., vol. R-35, N°5, Dec. 1986, pp. 544–558.

[PAG 88] L.B. Page, J.E. Perry, “An algorithm for fault-tree probabilities using the factoring theorem”, Microelectron. Reliab., vol. 28, N°2, pp. 273–286.

[PAG 80] A. Pages, M. Gondran, Fiabilité des Systèmes, Eyrolles, Paris, 1980.

[PAR 86] J.W. Park, G.M. Clark, “A computational algorithm for reliability bounds in probabilistic design”, IEEE Trans. on Reliab., vol. R-35, N°1, April, 1986, pp. 30–31.

[**PAN 88**] Z.-J. Pan, Y.-C. Tai, "Variance importance of system components by Monte Carlo", IEEE Trans. on Reliab., vol. 37, N°4, Oct. 1988, pp. 421–423.

[**PHA 88**] H. Pham, J. Upadhyaaya, "The efficiency of computing the reliability of k-out-of-n systems", IEEE Trans. on Reliab., vol. R-37, N°5, Dec. 1988, pp. 521–523.

[**POS 70**] V. Postelnicu, "Nondichotomic multi-component structures", Bull. Math. Soc. Sci. Math. Roumanie, vol. 14 (62), N°2, 1970, pp. 209–217.

[**POW 74**] G.J. Powers, F.C. Tompkins, "Fault tree synthesis for chemical processes", AIChE Journal, vol. 20, N°2, March 1974, pp. 376–387.

[**POW 75**] G.J. Powers, F.C. Tompkins, S.A. Lapp, "A safety simulation language for chemical processes: a procedure for fault tree systhesis", in [ESAR 75], pp. 57–75.

[**RAC 76**] L.E. Rackley, "System safety handbook for preparation of F-16 fault tree analysis", General Dynamics, Fort Worth Division, 16PR361, Aug. 1976

[**RAS 78**] D.M. Rasmuson, N.H. Marshall, "FATRAM-A core efficient cut-set algorithm", IEEE Trans. on Reliab., vol. R-27, N°4, Oct. 1978, pp. 250–253.

[**RAU 93**] A. Rauzy, "New algorithms for fault trees analysis", Reliab. Eng. Syst. Safety, vol. 40, 1993, pp. 203–211.

[**REA 75**] Reactor safety study. An assessment of accident risk in US commercial nuclear power plants, WASH-1400 (NUREG 75/014), US Nuclear Regulatory Commission, Oct. 1975 (Appendix II).

[**ROS 75**] A. Rosenthal, "A computer scientist looks at reliability computations", in Reliability and Fault Tree Analysis, SIAM, Philadelphia, 1975, pp. 133–152.

[**ROS 79**] A. Rosenthal, "Approaches to comparing cut-set enumeration algorithms", IEEE Trans. on Reliab., vol. R-28, N°1, April 1979, pp. 62–65.

[**ROS 79**] S.M. Ross, "Multivalued state component systems", The Annals of Probability, vol. 7, N°2, 1979, pp. 379–383.

[**RUS 83**] A.M. Rushdi, "How to hand-check a symbolic reliability expression", IEEE Trans. on Reliab., vol. R-32, N°5, Dec. 1983, pp. 402–408.

[RUS 85] A.M. Rushdi, "Uncertainty analysis of fault-tree outputs", IEEE Trans. on Reliab., vol. R-34, N°5, Dec. 1985, pp. 458–462.

[SAL 80] S.L. Salem, G. Apostolakis, "The Cat methodology for fault-tree construction", in [ASTO 80], pp. 109–128.

[SAS 76] M. Sasaki, H. Hayashi, "Confidence limits for redundant-system availability", IEEE Trans. on Reliab., vol. R-25, N°1, April 1976, pp. 41–42.

[SCH 84] W.G. Schneeweiss, "Disjoint Boolean products via Shannon's expansion", IEEE Trans. on Reliab., vol. R-33, N°4, Oct. 1984, pp. 329–332.

[SCH 90] W.G. Schneeweiss, "Fast fault-tree evaluation for many sets of input data", IEEE Trans. on Reliab., vol. R-39, N°3, Aug. 1990, pp. 296–300.

[SCH 00] W.G. Schneeweiss, The fault tree method, LiLoLe-Verlag, 1999.

[SHE 90] K. Shen, M. Xie, "On the increase of system reliability by parallele redundancy", IEEE Trans. on Reliab., vol. R-39, N°5, Dec. 1990, pp. 607–612.

[SHI 85] D.R. Shier, D.E. Whited, "Algorithms for genarating minimal cutsets by inversion", IEEE Trans. on Reliab., vol. R-34, N°4, Oct. 1985, pp. 314–319.

[SLA 70] J.R. Slagle, C.-L. Chang, R.C.T. Lee, "A new algorithm for generating prime implicants", IEEE Trans. on Computers, vol. C-19, N°4, April 1970, pp. 304–310.

[STE 81] K. Stecher, "Evaluation of large fault-trees with repeated events using an efficient bottom-up algorithm", IEEE Trans. on Reliab, vol. R-35, N°1, April 1981, pp. 51–58.

[TAK 82] K. Takaragi, R. Sasaki, S. Shingai, "An improved moment-matching algorithm for evaluating top-event probability bounds", IEEE Trans. on Reliab., vol. R-31, N°1, April 1982, pp. 45–48.

[TAK 83] K. Takaragi, R. Sasaki, S. Shingai, "An algorithm for obtainig simplified prime implicant sets in fault tree and event tree analysis", IEEE Trans. on Reliab., vol. R-32, N°4, Oct. 1983, pp. 386–389.

[VES 70] W.E. Vesely, "A time-dependent methodologie for fault tree evaluation", Nucl. Eng. Des., 13, 1970, pp. 337–360.

[**VES 81**] W.E. Vesely, F.F. Goldberg, N.H. Roberts, D.F. Haasl, Fault Tree Handbook, US Nuclear Regulatory Commission, 1981, Washington DC.

[**VIL 88**] A. Villemeur, Sûreté de Fonctionnement des Systèmes Industriels, Ed. Eyrolles, 1988, Paris.

[**WHE 77**] D.B. Wheeler, J.S. Hsuan, R.R. Duersch, G.M. Roe, "Fault tree analysis using bit manipulation", IEEE Trans. on Reliab., vol. R-26, N°2, June 1977, pp. 95–99.

[**WIL 85**] J.M. Wilson, "Modularizing and minimizing fault trees", IEEE Trans. on Reliab., vol. R-34, N°4, Oct. 1985, pp. 320–322.

[**WIL 90**] J.M. Wilson, "An improved minimizing algorithm for sum of disjoint products", IEEE Trans. on Reliab., vol. R-39, N°1, April 1990, pp. 42–45.

[**WLK 05**] A. Wilson, N. Limnios, S. Keffer-McNulty, Y. Arminjo (eds.), Modern Statistical and Mathematical Methods in Reliability, World Scientific, Singapore, 2005.

[**WOO 85**] A.P. Wood, "Multistate block diagrams and fault trees", IEEE Trans. Reliability, vol. R-34, N°3, Aug. 1985, pp. 236–240.

[**WOR 81**] R.B. Worrell, D.W. Stack, B.L. Hulme, "Prime implicants of noncoherent fault tree", IEEE Trans. on Reliab., vol. R-30, N°2, June 1981, pp. 98–100.

[**XIZ 84**] H. Xizhi, "The generic method of the multistate fault tree analysis", Microelectron. Reliab., vol. 24, N°4, 1984, pp. 617–622.

[**ZHA 85**] O. Zhang, O. Mei, "Element importance and system failure frequency of a 2-state system", IEEE Trans. on Reliab., vol. R-34, N°4, Dec. 1985, pp. 308–313.

[**ZHA 87**] Q. Zhang, Q. Mei, "Reliability analysis for a real non-coherent system", IEEE Trans. on Reliab., vol. R-36, N°4, Oct. 1987, pp. 436–439.

[**ZIA 86**] R. Ziani, "Vérification des objectifs de disponibilité et de maintenabilité des systèmes complexes modélisés par leurs ensembles minimaux", Thesis, University of Compiègne, GI, Dec. 1986.

[VES 81] W.E. Vesely, F.F. Goldberg, N.H. Roberts, D.F. Haasl, Fault Tree Handbook, US Nuclear Regulatory Commission, USA, Washington DC.

[VIL 88] A. Villemeur, Sûreté de Fonctionnement des Systèmes Industriels, Ed. Eyrolles, 1988, Paris.

[WHE 77] D.B. Wheeler, J.S. Hsuan, R.R. Duersch, G.M. Roe, "Fault tree analysis using bit manipulation", IEEE Trans. on Reliab., vol. R-26, N°2, June 1977, pp. 95-99.

[WIL 85] J.M. Wilson, "Modularizing and minimizing fault trees", IEEE Trans. on Reliab., vol. R-34, N°4, Oct. 1985, pp. 320-322.

[WIL 90] J.M. Wilson, "An improved minimizing algorithm for sum of disjoint products", IEEE Trans. on Reliab., vol. R-39, N°1, April 1990, pp. 42-45.

[WILK 06] A. Wilson, N. Limnios, S. Keller-McNulty, Y. Armijo, Modern Statistical and Mathematical Methods in Reliability, World Scientific, Singapore, 2005.

[WOO 85] A.P. Wood, "Multistate block diagrams and fault trees", IEEE Trans. Reliability, vol. R-34, N°3, Aug. 1985, pp. 236-240.

[WOR 81] R.B. Worrell, D.W. Stack, B.L. Hulme, "Prime implicants of noncoherent fault trees", IEEE Trans. on Reliab., vol. R-30, N°2, June 1981, pp. 98-100.

[XIZ 82] H. Xizhi, "The generic method of the multistate fault tree analysis", Microelectron. Reliab., vol. 24, N°4, 1984, pp. [illegible]

[ZHA 85] Q. Zhang, Q. Mei, "Element importance and system failure frequency of a 2-state system", IEEE Trans. on Reliab., vol. R-34, N°4, Oct. 1985, pp. [illegible]

[ZHA 87] Q. Zhang, Q. Mei, "Reliability analysis for a real non-coherent system", IEEE Trans. on Reliab., vol. R-36, N°4, Oct. 1987, pp. [illegible]

[ZHA 89] [illegible] "[illegible] des objectifs de sûreté de [illegible] des systèmes [illegible]", Thèse [illegible] de Compiègne, UTC, 1989.

Index

RECEIVED
FEB 19 2008